U0904063

都会桃源

销 信 中 心 设 计

主编 孔新民

中国林业出版社

序

接到孔新民主编的邀请，要我为本书写些文字，我仔细翻看了书中的作品，发现有许多项目是出自台湾设计师之手。这让我想起了去年11月份，我与孔主编等几位设计师朋友去台湾访问的经历。我们拜访了数位台湾非常优秀的设计师，参观了他们的公司及作品，其中大多为楼盘的销售中心及示范单位，看了台湾同行的作品，大家都感慨良多，颇受启发，并展开了热烈的讨论。这次孔主编邀我来写这本书的序，猜想大概是想再续以前那个意犹未尽的话题吧。

初学设计的年代，能够读到的专业书籍很少，偶尔买到一本台湾的《当代设计》便如获至宝，爱不释手。感叹于台湾室内设计作品的精致与意境。十几年后的今天对于设计项目我已阅历无数，而当真正对台湾设计师的作品身处其中时，当初的感受却丝毫没有减少，现场的感觉比书上看到的还要好。

谭精忠是我最喜欢的台湾设计师之一。不只是因为他设计手法的纯熟及设计经营的成功，我更感动于他对当代艺术品的执着热爱，并与自己的设计作品完美结合，给世人展示了设计与当代艺术结合的惊艳之美。在提升设计作品的精神境界的同时，做着自己喜爱的事，从而不知疲倦，这样的工作状态不是我们大家都梦寐以求的吗？参观谭先生的工作室，简直是一个小型的当代艺术博物馆，好多我们在艺术杂志上经常看到的艺术家作品在这里都可以看到真迹，多到不得不堆放在空间的每一个角落，到了这里我才认识到在谭先生作品中看到的如在高雄“都厅苑”接待中心出现的牟柏岩的《天空》树脂雕塑，胡栋民的《迎面》不透钢雕塑；“市政厅”接待中心中席德进、朱德群等当代大师的作品以及“国泰天母”接待中心示范单位中出现的“树灯”等，我看起来很眼熟的艺术作品只是谭先生个人收藏中的冰山一角而已。艺术收藏与设计的结合也是本人非常感兴趣的方向，所以对谭先生的作为颇为敬佩。

来到王玉麟先生的“岳泰峰范”，印象最为深刻是其无与伦比的精致，闹中取静的环境营造，空间尺度的准确拿捏，材料表情的完美配搭，工艺节点的细致入微，结合高超的施工建造，其精致程度让我等一行人深感震撼，尤其是在其中示范单位的部分，更是无可挑剔。而那一刻正是整个示范区即将拆除的前一个小时。同时，王先生也是一个可以演绎不同设计表情的百变设计师。“岳泰峰范”、丽宝“city one”

序

以及另外一个我亲历过的“展悦”接待中心都有着不同的表情，但都一样得到完美的展现。

程绍正韬是我认识的台湾设计师中最富激情的一位，就象他的座驾，那部亮咖啡色“911 carrera”，动力澎湃却又极其经典。他可以激情四溢地演绎变幻莫测的建筑及室内空间，像一个空间魔术师，同时又精通中国古代典集，常聚三五知已，坐而论道，把酒言欢，颇具中国古代贤哲风范。他告诉我们他最热衷做的事就是把中国古代的文学、艺术作品中的情境以现代的设计手法演绎出来，以追寻古代先哲的人文精神。在本书收录的“AZ学生活”接待中心的设计案中，你可以感受到设计师在空间演绎的过程中给人们传递的具有东方文化精神的感受。想更多地解读程绍，我建议你去看一看他设计的位于台中的“私房”泰餐厅，你的感受会很丰富。如果能够约到他本人，在户外的草亭里，小酌一番，不失为难得一次思想盛宴。

我们拜访过的几位台湾设计师团队规模都不是很大，少则几人，多则二十几人，但却各有特色，设计得很“自我”。如拜访过却未及多加描述的陆希杰先生、张文信先生、黄书恒先生等。我至今难以忘怀黄书恒先生的那间香烟缭绕的标准高技派风格的办公室……

在本书的作品中，我也看到了从不缺乏创意的大陆设计师的作品的愈发精致，“精致”一直是我个人的设计追求，所以会更关注这样的变化，“精致”的作品来源于成熟的设计、完善的市场配套，以及高超专业的施工建造，达到这样的成果是大陆装饰市场整体水平快速提升体现，让人欣喜。也给我个人增加了学习与进步的动力。

我相信每一个不同的经历都是一次学习和进步的机会，正是抱着这样的一种学习心态，我才敢斗胆接受主编的邀请，为此书作“序”，自知无法胜任，故请同道原谅本人的才疏学浅与信口雌黄，如能抛砖引玉，换得一些批评指正，达到交流学习的目的，本人将非常感激。

于　强

2011年3月21日

CONTENTS

CONTENTS

FUXING NATIONAL CITY SALES OFFICE IN WUHAN

武汉福星国际城售楼处

本案的概念设计定位为年轻活力、时尚休闲的商业化空间，与福星国际城整体规划理念相一致。在这里，各种几何元素既完美融合又各自独立，黑白红黄强烈的色彩对比碰撞出独特的空间感受。

作为一个全开放式的功能空间，本案的功能划分以及流线设置由发光天花造型以及地面与之相对应的深灰色哑光砖来完成。由不锈钢条，阻燃金属砂组合成的超大尺度的发光天花造型，以相对较低的标高，强化了交通空间，并分隔出不同的功能区域，同时也是整个销售中心重要的泛光源，塑造了柔和现代的空间基调。

入口左边梯形区域独立设置圆形接待台，醒目的红色拉开这个时尚活力空间序幕。右边洽谈区，白色的沙发，白色发光斜立柱，与原建筑黑色直立柱以及地面黑色高亮砖交相辉映，打破了空间沉闷气氛，并与建筑外广场支撑浮云餐厅斜柱相呼应。斜立的绿植延续了设计语汇，也为洽谈区增添了些许生气。

灯光设计也成为本案造型中的一部分，从天花条形灯罩，到洽谈区发光的斜立柱；从接待台、水吧台以及玻璃丝印墙面的地面暗藏环绕灯带到旋转楼梯内侧嵌入式条形灯带，从视觉上分隔区间，指引流线，也从细节上延续了空间设计语言。

工程名称：武汉福星国际城售楼处

坐落地点：湖北武汉

主持设计师：李益中、周伟栋

面　　积：1200 m^2

主要材料：阻燃金属砂、深灰色氟碳漆、暗红色氟碳漆、玻璃丝印图案、白色哑光砖、深灰色哑光砖、黑色高亮砖

工程造价：220万

竣工时间：2009.10

设计单位：深圳市派尚环境艺术设计有限公司

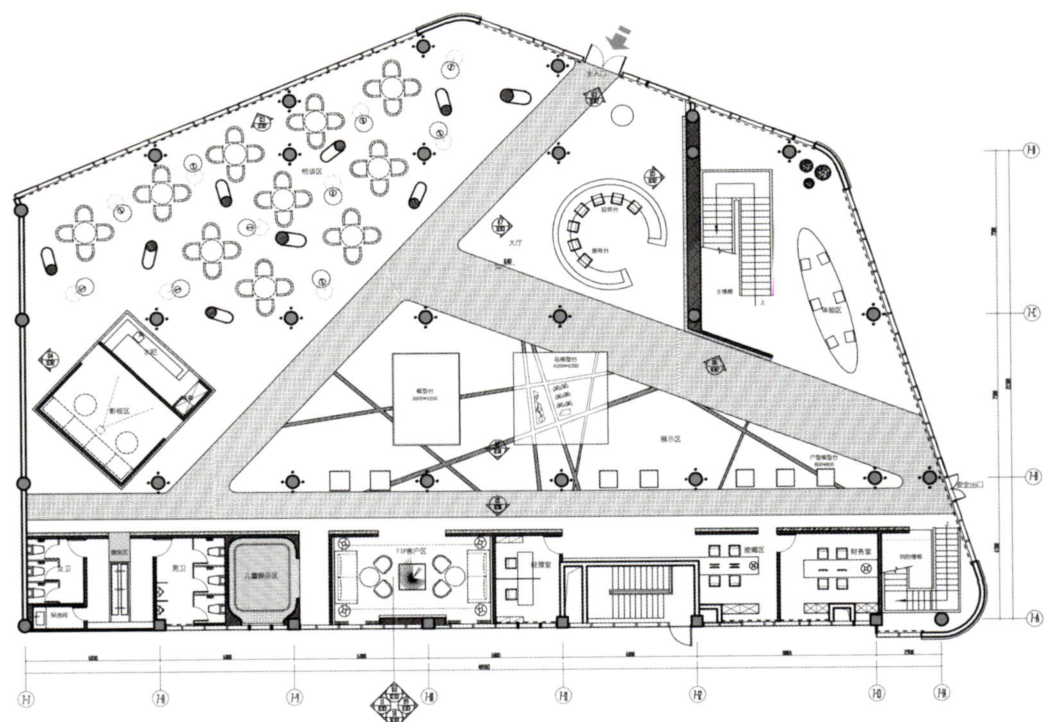

Conceptual design of the case is regarded as a young, vibrant, fashion commercial space, which is as same as the overall planning concept of International City of Fuxing. Here, variety of geometric elements not only fuses perfectly but also keeps independent. Intense collision with black, white, red and yellow gives visitors a unique feeling of space.

As a wide-open function space, the functional partition and the flow lines of the case are achieved by the luminous ceiling and dark gray ground matte tiles corresponding. Formed by the stainless steel, fire-retardant metal sand, the shape of large-scale luminous ceiling strengthen the transport space and separate out the different functional areas by a relatively low elevation.And it is also the important pan illuminant throughout the sales center, shaping a soft modern tone for the space.

On the left trapezoidal area of the entrance set a circular reception desk, the striking red pull off the fashion dynamic space. The negotiate area on the right which contains white sofa and white luminous oblique column stands with the straight, black columns and the black highlighted brick of the original building enhances beauty of each other's , breaks the dull atmosphere and produce a particular effect in the case of combination with the fastigiated column supporting FUYUN restaurant in the square outside the architecture, adding some more vigor to the discussion area.

Lighting design has also become part of the case modeling, from the linear ceiling lamp to the white luminous oblique column in discussion area; from reception desk, water bar and the hidden lights on the ground along the silk-screen glass wall to the embedded linear lamp on the inner side of the spiral stair, separates the space and gives directions visually. They extended the space design vocabulary in detail at the same time.

BINHAI WANKE CENTER

天津滨海万科中心

IN TIANJIN

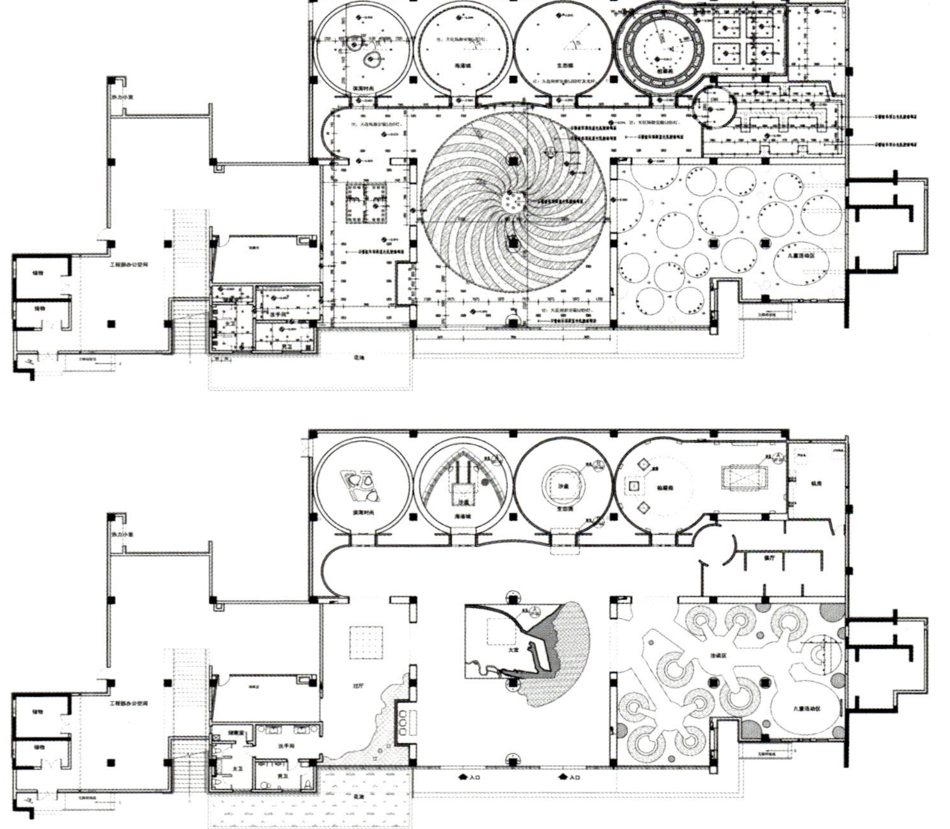

天津滨海万科中心是伴随着滨海新大区的建设及天津万科加大对滨海投资及品牌推广后孕育而生的产物，中心内集合了万科品牌推广、滨海项目的推介、楼盘项目洽谈等多种功能。设计师采用天、人、海的设计元素使整个空间完美的形成一个整体。

参观动线为：滨海新区区域介绍—万科品牌推广—滨海项目推介（滨海时尚）（天津港）（生态城）（柏翠园）—滨海摄影展—休息洽谈（含儿童活动区）。

滨海新区区域介绍（天），进入展厅15 m直径的螺旋形吊顶及滨海大区区域模型上下呼应占据了大厅的主要空间，螺旋形吊顶符号来源于银河旋臂，在旋臂之上星星点点的LED灯光将宇宙的浩瀚演绎得淋淋尽致，地面采用的是黑星石英石，石材上镶嵌的碎镜面完美的将顶面的LED灯光反射，模型采用防水材料制作，坐落在等比缩小的滨海海岸线之上，水穿过城市缓缓流入大海。光影设计在这个区域的使用也是设计时的重点之一，在天空旋臂的缝隙内6台高明度投影仪通过熔接处理完全覆盖于区域模型之上，从而营造出日月轮回的时空效果。电脑灯

工程名称：天津滨海万科中心
坐落地点：天津滨海万科金域蓝湾底商
设 计 师：张铁然
面　　积：1280 m²
设计时间：2010.05
竣工时间：2010.08
摄　　影：傅建勇
设计单位：北京谷雨建筑设计有限公司

则负责在项目推介时提示项目所在位置，配合背后宽幅投影做到同步演示介绍项目的作用。

万科品牌推广，由于此空间相对较小，因此在设计时设计师采用雕塑类装置的形式用最简单、直观的方式把万科在全国的所有项目一一罗列。

项目推介（人），此区域共分成4个展示仓，分别代表了万科即将在滨海推出的4个主要项目。滨海时尚，作为万科在滨海最高的住宅项目，设计师将“高”这个理念充分运用，墙面采用同项目建筑外观相近似的方形孔洞处理，中间采用超尺度建筑模型来体现建筑的特点，同时设计师还将中国传统的走马灯运用到建筑模型中，不同的场景及人物的千姿百态在建筑模型中穿流。天津港，站在这个展示仓中间仿佛站在一艘豪华游艇的甲板上俯瞰天津港的繁华与美丽。生态城，设计师借鉴了“阿凡达”中树精灵的符号游戏，在仿佛钟乳般的空间内，让人体会的是另一个空间的宁静与美丽。柏翠园——万科推出的高端住宅项目，设计师用3座19世纪30年代法国制造的纯铜座钟来衬托柏翠园的品质与收藏价值可谓相得益彰。在这个区域除了项目特点介绍外，设计师还将游戏的趣味运用到设计中，进入每个舱门的地面上设有密码，参观者必须按密码提示将两只手分别插入相对应的数字孔洞中才能打开舱门，从而使参观者真正的参与其中。

滨海摄影厅则为参观者展示着滨海及万科的过去、现在和未来。

洽谈区（海）为突出展厅的地域性，设计师采用海浪为设计原型，让参观者在高低起伏的波涛中体会项目带给自己的愉悦。

Wanke Center in Tianjin Binhai is born accompanied by the construction of Binhai New Area and Tianjin Wanke increasing investment and brand promotion, the center puts Wanke brand, Binhai projects promotion, real estate project negotiations and other features together. Designers use elements including sky, people, and sea to make the space form a perfect whole.

The viewing line: Binhai New Area Description——Wanke branding——Binhai project promotion (Binhai fashion) (Tianjin Port) (Eco-City)——Binhai Photography Exhibition- negotiation room (including children's activity area.)

Binhai New Area introduction (sky): Entering the exhibition, hall 15 meters in diameter and the spiral-shaped ceiling coastal region accounted for the regional model of the hall echoed up and down the main space. When designed this area, lighting effect is one of the focus point. 6 sets of high light projector completely

covered the processing model in the region fusion to create a space-time effect of the sun and the moon cycle in the gaps of the spiral arms in the sky.

Project promotion: This area contains four Exhibition Halls, representing four major projects launched by Wanke in Binhai. Fashion Binhai, as the highest Residential project for Wanke, makes full use of the concept of "high". The wall is decorated with square hole similar to the outline of construction. In the middle ,the characteristics of the architectural is reflected by an ultra-scale, while the traditional Chinese Running Horse Lantern principle is applied to the building models to make a circumstance that different scenes and characters of different styles in the Model flowing.

In Binhai photography hall, by watching the exhibition visitors know well about Wanke's past, present and future. Negotiation room (sea), to highlight the regional of this exhibition hall, designers use waves as prototype, making visitors enjoy the pleasure brought by the project which make people ups and downs in the waves.

滨海
悦变
摄影展
BinHai
Style photography
exhibition

VOYAGE FOR SALE

深圳大梅沙八十步海寓销售中心

位于海边的销售中心，销售的是欧式风情的商业街与度假公寓。赋予室内度假的悠闲气氛，并将景观与建筑引入室内。植物墙、室外石材与实木是主要材料，而西式装饰元素则用在灯具与家具等细节上，在度假的氛围中点缀出销售的欧式情调。

项目名称：深圳大梅沙八十步海寓销售中心
项目地址：深圳大梅沙
项目面积：600 m^2
竣工时间：2009
主要用材：非洲柚木地板、非洲柚木实木、砂岩
设计单位：于强室内设计师事务所

On offer in the seaside Sales Centre are commercial property units in the European-style shopping street and resort apartments. The Sales Centre is imparted with an ambiance of leisure vacation in its interior design, bringing the outdoor scenery indoor. The vertical plant-decorated wall and stone materials brought in from the outdoors together with the hardwood panels are the main building materials used. Western style décor elements are applied onto such details as the light fittings, furnishing, and so on, directing attention to the European flavour of the Sales Centre set in a holiday atmosphere.

QIUSHI E SALES OFFICE

秋实e景售楼处

秋实e景家园售楼处坐落于长春市繁荣路8号，它的建筑面积为650m²。在形式上它迎合了业主一贯坚持的时代性原则，把信息技术与生活艺术融合在一起，既体现时代精神，又要彰显人文思想，是物质与精神在物化形式上的高度统一。

在此前提下，设计师以"e"的信息母语为起点，在计算机语言形式中抽取线性结构，并通过对现场空间的认识和理解，把功能场所分为明、暗两个空间区段，明空间为深度洽谈区，暗空间为体验观赏区，两个空间区段中由总服务台进行空间连接，在形式上起到了一定的支配作用，同时在接待上也满足了引导服务性功能。就线形的灯带划分而言，我们可以看到两个方向上的空间变化，一个是顺向的，一个是横向的，两个运动特性在同一个空间中产生对撞，使进入者第一感受就是时间的流失，及生命感的自我重塑。在此，语言已无力与空间抗衡，只有自身亲为，才能满足心灵那份久违了的精神空缺。

空间的本质是用来体验的，而不是说教的，体验是建立在对建筑实体围合下的"虚"的存在，而"虚"是形式背后的真实属性，设计师正是本着这样一种理念，在线性的重构中诠释了内心对"e"时代生活的场景描绘，也抒发了人文情怀下业主对客户的无限关爱之情。

在本空间中材料的运用与表达始终贯穿于设计风格之中，条形的木质饰面板由线性转换到面，再由面与条形灯带形成图底关系，这一宽一窄，一虚一实，一暗一明从墙壁过度到顶棚使材料的肌理在灯光的映衬下形成空间界面的自由转换，它大大增强了空间意象由触觉感向视觉感转换的这一知觉过程。为了使木板这种肌理看上去更加细腻些，特在水景处增加了一处毛石劈面剪影山形墙体，它与玻璃、白地砖、木质形成强烈的质感对比，同时它也寓意自然要素在理性中得以升华，道家的禅意山水境界在此得到

了充分的体现。

售楼处自投入使用之后，同我们设计之初所设想的效果完全一致，它不但得到了业主充分的肯定，也为前来咨询的顾客提供了一个超出想象的空间感受，整体空间虽然是由抽象的几何形体建构的，但整个空间所传达的知觉意象是科技与人文的思想。这种思想理念在形式的契合中依据设计师对抽象形式的准确把握以简约、洗练的艺术表现形式创造出能够让人兴奋不已的感性空间，这也是设计师通过秋实e景这一工程的实践向人们展现了精神与物质、感性与理性间的完美结合。

Qiu sales offices located in Prosperity Road, Changchun City, No. 8, its building area is 650 sqm. It caters to the owners who consistent adherence to the principles of the times, putting the information technology and art of living together, not only embodies the spirit of the times, but also highlight the human thinking. It is high degree of unity of material and spiritual.

Under this premise, The designers use letter "e" as the starting point, in the form of computer language to extract linear structure, Through on-site awareness and understanding of space, the functional areas are divided into light and dark sections of two space by the designers . Discussion area is the light space, dark space for the experience of viewing area, two sections of space connected by the main desk which plays a dominant role and also has the guidance function when reception is needed. As far as the partition of the linear lights is concerned, we can see the spatial variation in both directions. One is forward, another is horizontal, the two movement characteristics collide in the same space, so that the first feeling to visitors is the loss of time, and life sense of self-remodeling.

To experience the essence of space, rather than didactic, Experience is based on the "virtual" existence under the entity construction, and "virtual" is the true characteristic behind the form, insisting such ideas, designers reconstruct the linear to

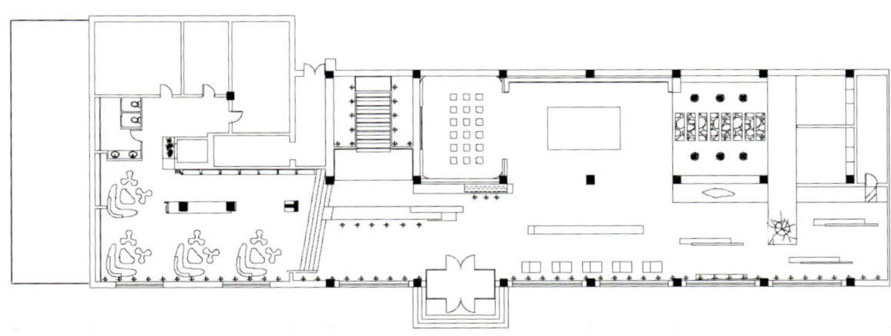

interpret what is the life of " e " era in their hearts, and also to express unlimited caring to customers.

In this space, the use of materials and expression goes throughout the design style, bar-shaped wooden panels transform from line to plane, and the planes would be used as the background by the linear lights.

A wide one, a narrow one; a virtual one, a real one; a dark one, a bright one, these comparisons last from the wall to the ceiling, makes the texture of the material in the light against the background create the free space interface conversion. It greatly enhances the perceptual process that space images change from the tactile sensation to the visual. In order to make the texture of wood look more delicate, designers particularly add a silhouette of Gable wall in the waterscape, it forms a strong contrast with glass, white brick, wood texture, and also symbolizes that the natural elements are in the sublimation of reason, Zen in this has been fully reflected.

Since the sales office has been put into use, its effect has met what we wanted. It not only has been fully affirmed by the owners, but also provides a feeling beyond the imagination for the customers who come for the consultation. Although the overall space is constructed by abstract geometry, but the idea conveyed by the image of space reflects technology and humanities.

PIXEL IN BEIJING
北京像素

超高层综合住宅开发楼盘的样板间兼售楼处。与大约一万户跃层式户型组成的住栋设计一致，"像素"灵活运用于外装及内装的多个部分。将吊顶满布和纸做成的照明灯具，实现了同种设施中所追求的冲击力。样板间分为"流行"等4种不同主题的类型。

北京像素销售中心是在原有销售中心的基础上，重新修改使用的"北京像素项目"整体"细胞"概念这一风格而建成。

外观

外观是由多个各自独立的管堆积成一体构成。

管分为白色、浅灰色、深灰色、黑色4种，各色管长度不一，重叠累积，从而产生凹凸感。

销售中心大厅

销售中心大厅是用做"北京像素项目"整体销售的场所，包括洽谈室、贵宾接待室等等，是销售中心非常重要的组成部分。包括来宾入口在内，整个大厅设计都力求追求一种拥有和洽谈室相称的豪华感和震撼力的空间。

内部装修也意识"北京像素项目"整体细胞观念，墙壁部分累积400×400的正方形木箱，给予各个木箱不同的长度和色泽，创造出凸凹层次感。顶棚部分是由400×400的边框状的日本纸构成的筒从顶棚垂钓下来，起到照明作用。

地板部分注意内、外部分的一体性，是将室外园林风格直接引入室内这样一种风格。

样板间

本项目样板间的设计围绕以下主题展开："流行"（POP）。这些主题是依据不同生活方式的人而定的，可以生动表现出主人的性格。这些生活方式正是北京这样建立在先进、多样化都市上的，自由新生活的代表。

"流行"是平层复式结构的样板间。其中4.5 m高的挑空是这两种样板间的特点。我们根据各户型的不同，尽可能的把竖井做大。

我们把"流行"假设为一个20多岁年轻前卫的女服装设计师的家。我们假想她喜欢白色和红色，还喜欢收集这两色的公仔和帽子。因此我们做内装时，包括家具在内都使用了红白两色，或者它们的中间色浅红色作为基准色，房间里各种地方的家具上都摆放着主人收藏的收藏品。天井的一面侧墙和顶棚贴着巨大的镜子，可以映出置身在众多收藏品中的主人的身影，充分的满足了她自恋的心理。

项目名称：北京像素样板间
项目地点：中国北京市
设计周期：2009.02-05
施工周期：2009.05-12
竣工日期：2009.12
规　　模：地上4层
建筑面积：3000 m^2
设 计 师：迫庆一郎、最上有世、栗本贤一
设计单位：SAKO建筑设计工社

It contains sample rooms and the sales office of the Comprehensive high-rise residential development. The same as about ten thousand Duplex-style living units, "Pixel" is widely used in exterior and interior of the multiple parts. Model rooms contain four different styles including "POP". Beijing pixel sales center which is on the basis of the original sales center is built by re-modifying "cell" concept proposed in Beijing Pixel Project.

Appearance: Appearance is composed by some Independent tubes, including white, light gray, dark gray, black four colors. These tubes have different length, resulting in convex and concave sense.

Sales center hall is used as the overall sales place for "Beijing-pixel project ", including negotiating room, VIP room, etc. It is a very important part for sales center, including the entrance, design of the entire hall has sought to make the place luxury and shocking like negotiating room.

Interior decoration: to meet the "cell" concept proposed in "Beijing-pixel project ", the wall accumulates 400 × 400 square wooden boxes. Each wooden box is given a different length and color, to create a convex and concave sense.

Floor: making inner and outer parts together is a style that introduces Outdoor garden style to inner.

Sample room: Conception of the model room is about "POP". The theme is based on different lifestyles which can vividly show the owner's personality. Different lifestyles are representative of a new life in Beijing (an advanced and diverse city).

"POP" is a sample room of flat double layer structure. And the characteristic of 4.5 meters high can be picked between the two samples. Depending on different apartments, we make the shaft as big as possible.

We assume that "POP" is a house owned by a 20-year-old female fashion designer. We assume she likes white and red, and also likes to collect two-color hats and dolls. So when we do interior, including furniture, we use red and white, or light red color between them as the base color, various parts of the room are placed the owner's collections, you can find them easily on the furniture. One side of the patio and ceiling are decorated by huge mirror, which reflects the owner standing in the middle of her collections, sufficient to satisfy her narcissism.

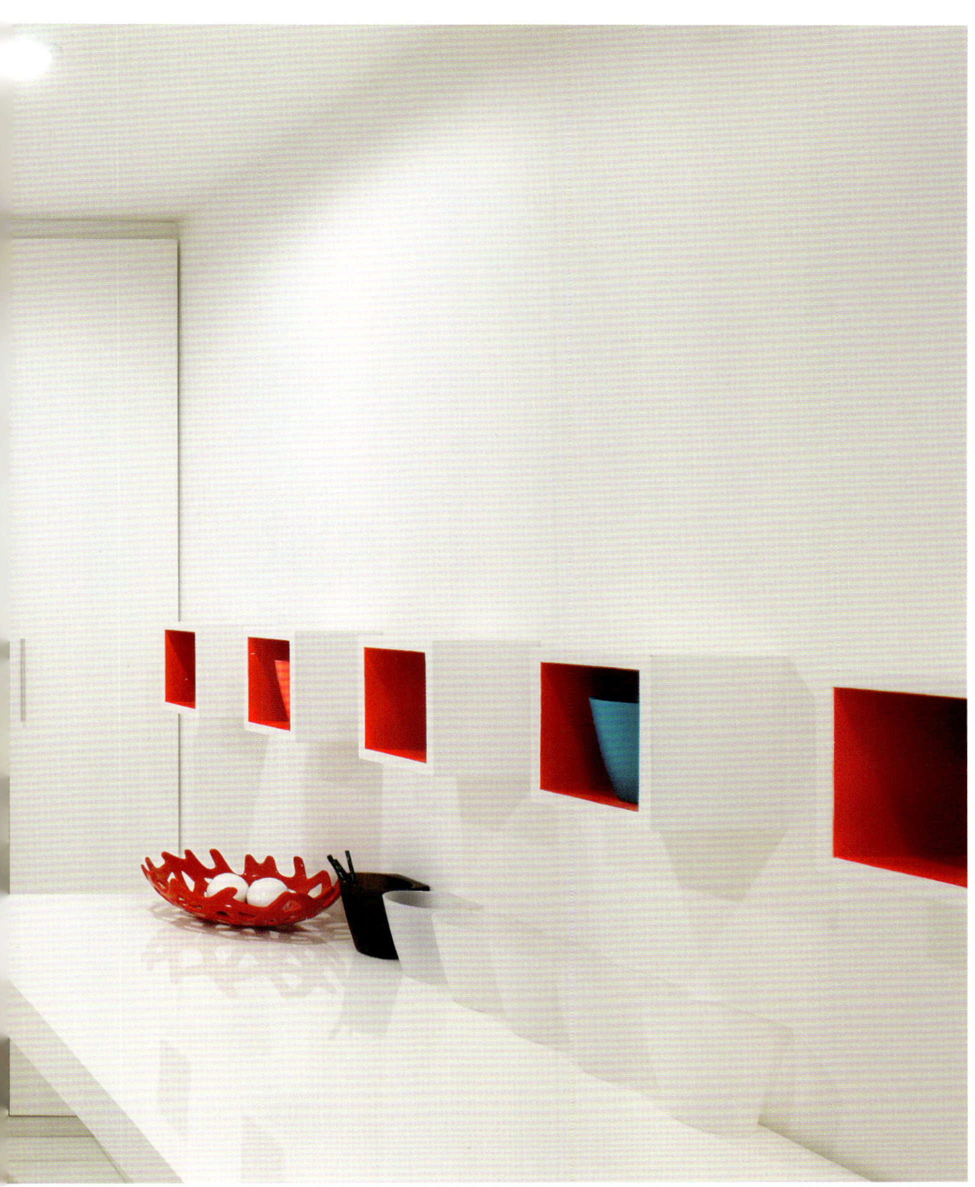

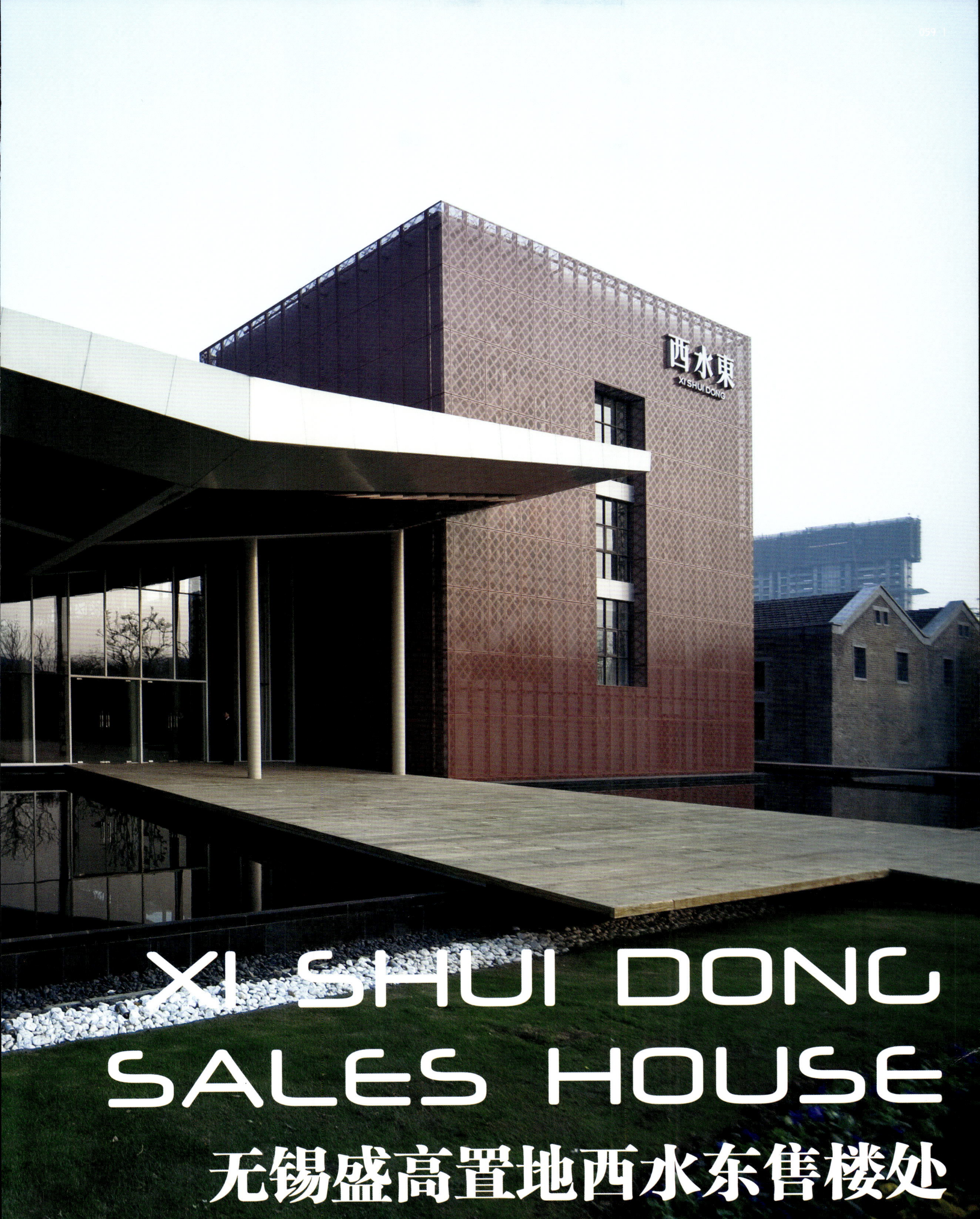

XI SHUI DONG SALES HOUSE

无锡盛高置地西水东售楼处

项目启动之前正值中国无锡市政府力图将其南边的一大块土地进行全新规划改造，同时作为开发商的香港盛高置地邀请全球知名的建筑事务所SOM和PTW以及Kokaistudios共同合作，将该地块打造成全新的综合高端住宅、商业和文化中心的城市新地标。由于项目基地存在大片历史保护工业遗迹和历史建筑，又夹杂部分后期新建建筑，因而建筑设计必须考虑到新旧建筑的协调及其与周边环境的融合。基于这个设计理念，Kokaistudios被邀请作为这个新开发项目的标志性建筑即西水东售楼处的总设计师。2009年末该售楼处顺利建成并获得广泛认可，整个住宅项目的销售盛况空前且在中国房产业内设立了全新的销售记录。

为了将整体项目的总体理念融入到这个售楼处的设计中，Kokaistudios选择对3个基地现存的历史仓库进行改造，同时新建一栋顶部为一对飞翔的翅膀造型的透光屋顶构成的现代建筑，翱翔展翅的羽翼将这组建筑群围合起来，形成一个全新的整体群落。售楼处整体的效果给人留下强烈的现代感，成为了整个项目的标志。透光屋顶通过三维设计在视觉上得以扩张，为来访的客人营造强烈的视觉透视效果。整个售楼处的建筑体系透过前部的水池得以强调和完整，水池可以收集天然雨水，同时点亮建筑本身，并为公众提供了有趣的体验感。

透光屋顶的设计不仅用以衔接现存的几个仓库，同时也为售楼处室内提供了极高的层高空间，也强化了接待大厅的视觉效果。天窗的设计辅以轻盈的木质天花装饰为室内营造了日夜各有变化的光线效果，也使得屋顶的整体感觉更轻盈，给人更大的愉悦感。在视觉上平衡横向的羽翼结构的是其一侧的一栋现存工业塔楼，我们将外立面进行重新改造设计，使其能与整个售楼处的风格相融合，并为整个售楼处增添了一处新的标志。

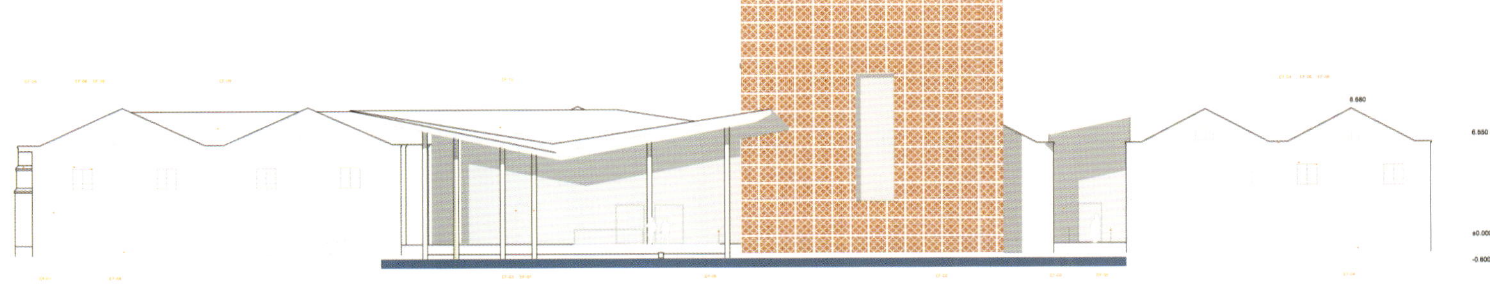

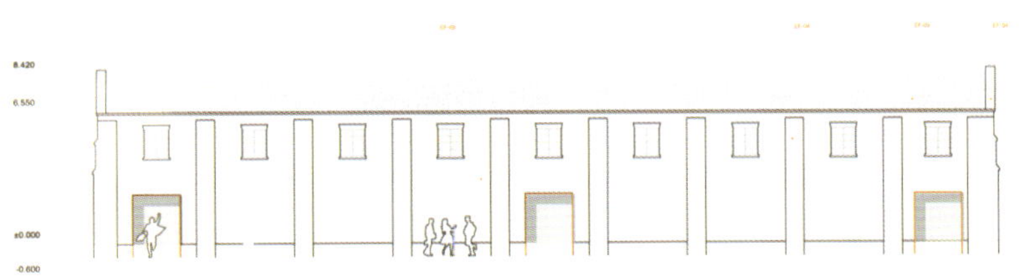

项目名称：无锡西水东售楼处
地　　点：中国江苏省无锡市
面　　积：3437 m^2
设计时间：2009.03
竣工时间：2009.10
客　　户：盛高置地
设计团队：Filippo Gabbiani, Andrea Destefanis, 李伟、张怡、宋庆、俞峰
摄　　影：PANDA of VA-PHOTO

When the large area located to the south of the Chinese city of Wuxi was selected by the city to be reconverted and re-qualified the Hong Kong based developer SP-Gland invited a selected number of international firms including SOM, PTW and Kokaistudios, to cooperate and transform it into a new residential, commercial and cultural center of the city. As the site was mostly occupied by a heritage factories of the city, large portions of historical buildings have been maintained and integrated together with the new contemporary ones, setting a strategy of harmonious development between old, new architecture and the surrounding environment. Being this the vision proposed by our team all over the project we have been invited to design the iconic project of the sales center of the new development. The success of this strategy was confirmed at the opening of the center in late 2009 when the entire project has been sold out setting a new sales record for the Chinese market.

Encapsulating the spirit of the project in one sales center, the choice was made to restore three existing warehouses of the previous factory and integrate them with a new, modern architectural building topped by a light roof composed by two flying wings that embrace and complete them. The effect is an utterly compelling contemporary building that serves as the symbol of the entire project. The effect of this light roof is amplified by the tri-dimensional design conceived by setting the sloping to the front and creating a strong perspective effect to the incoming visitors. The entire sys-

tem determined by this new light roof, is supported on the front part of the building by the creation of a water pond that collects the rain waters, light up the building and create an experience for the public.

The light roof has been conceived to connect the existing warehouses and create a new indoor space with high ceiling height, impressive reception lobby of the sales center. The lightness of the roof and the comfort for the people is amplified by the design of skylights that together with a special light wooden ceiling create different pleasant lighting effects during day and night. Balancing the horizontal dynamic effect of the flying wings is an existing auxiliary tower, reconfigured by Kokaistudios by adding a new skin to the existing industrial structure. This volume previously a disordered addition become integrated part of the iconic architecture, a new symbolic entity of the new life of this project.

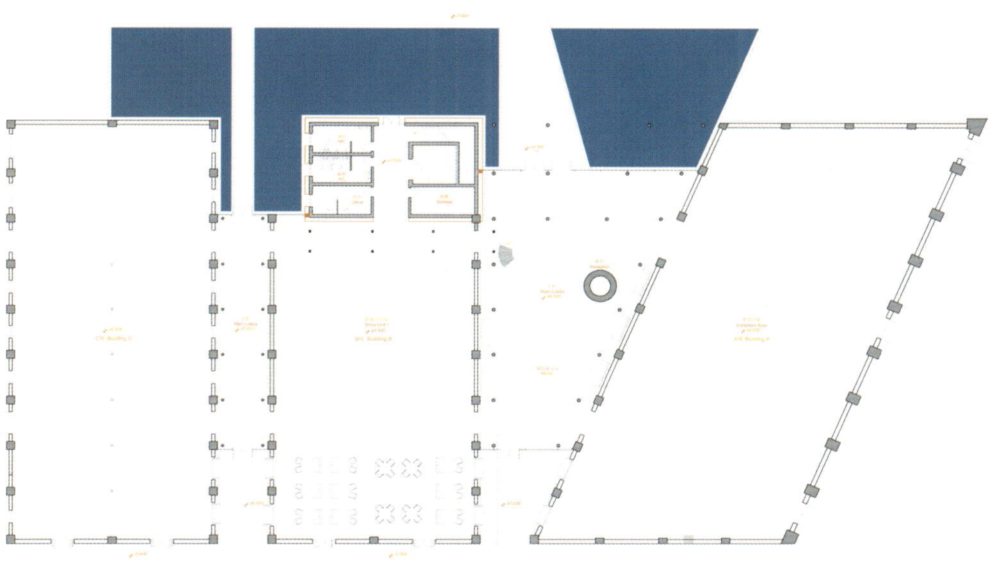

TAIYANG XINGCHENG SALES CENTER

北京太阳星城售楼中心

线性强调的是一种规矩，一种容易被识别的属性。正如我们所习惯的室内空间一样：一体六面，墙需直，地要平。从左到右，从上到下规矩的直白，这就是线性建筑给我们的感受。而伴随着信息科技发展带来的非线性空间强调的是人在空间里的自由度和对空间功能丰富的模糊性。曲线建筑和曲线空间，建筑界内称之为"非线性建筑"。它是根据非线性理论而来的，线性即一次函数，图像为一条直线，而非线性是指自变量与变量之间不成线性关系，成曲线或者抛物线方向，即非直线。曲线空间的流动和渗透性，造成空间自身的混沌性、模糊性、非标准性和丰富性。在这样的空间里活动，必然会产生比"方盒子"建筑更丰富的感受和多维的体验空间，实际上是扩展了空间的维度。

这次北京太阳星城销售中心的室内设计就进行了这种非线性设计的尝试。这种尝试也是经过对未来楼盘业主的调研后而制定的，伴随着科技社会成长起来的人们，生活观和价值观也与以往社会形态下的人们有了很大的区别。世界前所未有的小，也开始空前的丰富。人们追求更优质的生活品质和开始追捧不断变化的时尚，他们厌倦直白乏味的产品，定期推陈出新的高科技产品是他们喜爱的对象。他们更倾向于选择一种能丰富内心感受和体验的事物，彰显他们独特个性的产品。而这样一种非常规，非标准的需求，将会在现代的社会里被持续需求。

北京太阳星城住宅区由金星园、木星园、水星园、火星园和土星园组成。为其新设计的销售中心的室内设计概念同样来自于"星际空间"。这种概念从平面功能的设置到立体空间的表现贯穿始终。在销售中心里展示硕大的太阳星城沙盘一直是开发商需要的，设计师在仅三百多平方米的销售空间里解决了这个难题：把室内的地平面提高了半米，模型被置于观众脚下的玻璃地面内，客人随意的可以近距离了解喜欢的楼盘。抬高后的地面让设计师更放松的利用了不同的地面层降关系来布置其它平面功能的要求：洽谈区在黑暗的太空中，背景钻石突起的墙面似有温度的自转陨石；签约区被放在了类似陨石坑的地方；VIP区则漂浮在空中的飞行器里。室内的销售空间明显的被黑白划分成太空和星球，洗手间区域的设计则更加表现时空的转换和停止：达利那个挂在树上的钟表也被扭曲在天花上了。可惜的是因为其他原因，入口处的室外空间最终没有按原设计来实现，我们只能用设计图来感受设计师想表达室内外设计统一的观点，高耸环抱的入口造型从室内涌出室外，直指夜空，带给无限遐想。

太阳星城售楼中心的设计概念——"星际空间"运用非线性的设计元素，大胆分割和包围空间，彻底改变原本只是个方盒子的空间，不再是乏味的直，而是在营造一种似如宇宙的星际空间，空间于是具有了多种方式，从而也开启了无数的可能。

01 大厅立面图 ELEVATION
SCALE: 1/30

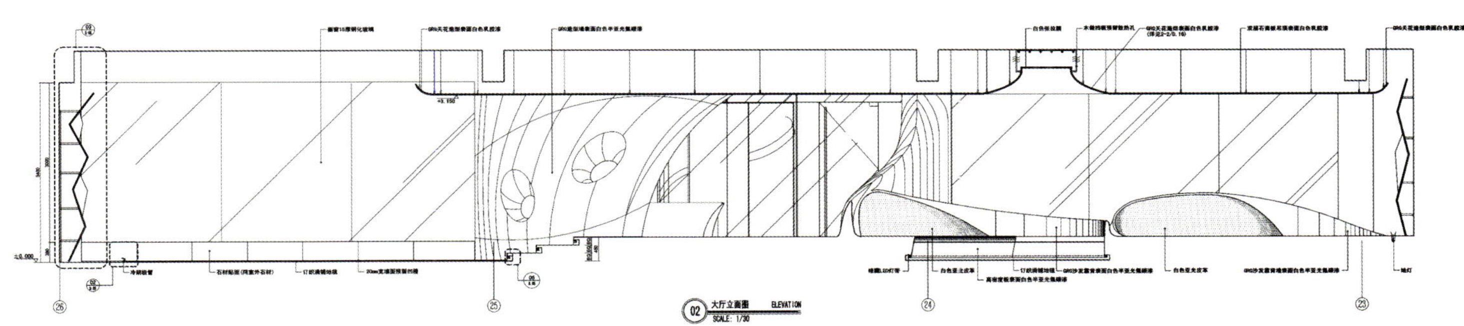

02 大厅立面图 ELEVATION
SCALE: 1/30

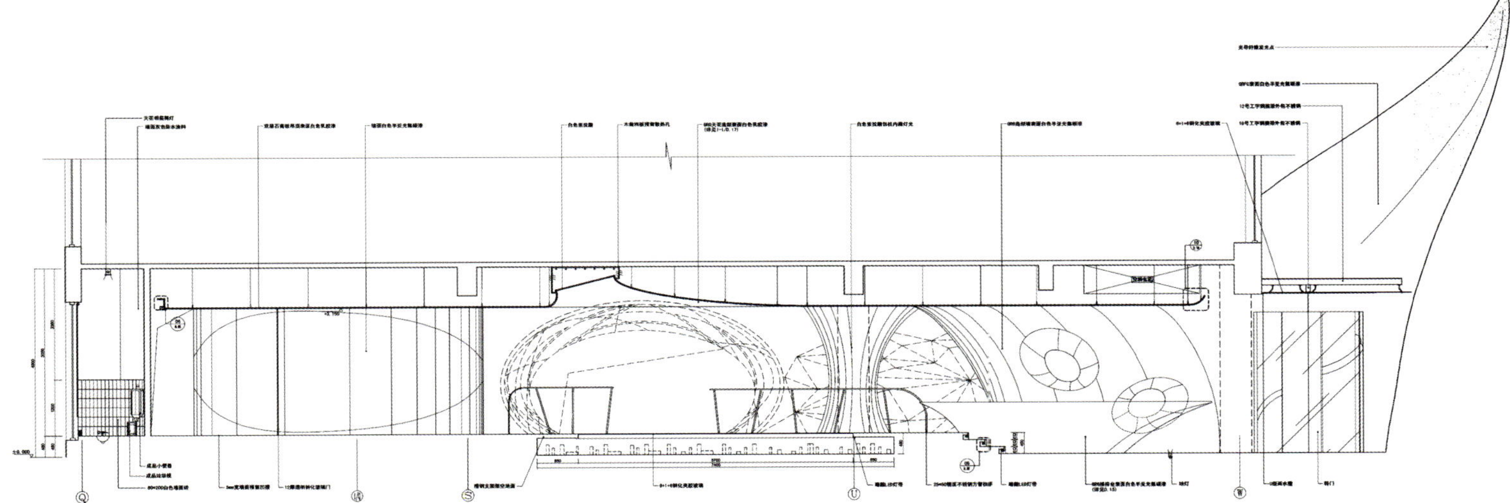

项目名称：北京太阳星城售楼中心
坐落地点：北京朝阳区太阳宫中路
开 发 商：北京冠城正业房地产开发有限公司
设计单位：北京立和空间设计事务所
施工单位：北京冠丰装饰装修有限公司
竣工时间：2009.7
设计总监：贾 立
设计管理：张 倩
设 计 师：张立超、张旭、张德勋、银旭

This case presents a typical nonlinear space. The case is mainly used for house selling and offices. The theme of the housing is "Interstellar Space", including "Venus Space", "Jupiter Space", "Mars Space", "Mercury Space" and "Saturn Space". According to the theme, the sales centre is designed from plane function to three-dimensional space.

Straight lines are hardly to be seen. Instead curves and irregular angles are widely used. White is used extensively. The light is mirrored on the self-leveling white floor, meanwhile, the smooth step line traffic and the ceilings set each other off beautifully. The designers imitate the interstellar space in which each function area is set and separated abstractly. In terms of space and the extension of thinking, the whole design is endowed with plenty of imagination and extends the space dimensions so that it makes a regular cube have countless possibilities.

The reception table and the cashier

are just opposite the entrance. The red huge cylinder is a comparatively closed cashier, which has certain privacy. However, contract signing area and VIP area are on each side of the entrance. Contract signing area is designed in a round hollow area, as if it seems like a huge meteorite pit, while the VIP area is more like an oval flying spacecraft, which makes the materials overlap just like a nest. The washing room in this case is full of Surrealism. Its front part is mainly white with mirrors in it. The interior part is mainly red which provides a contrast for the front part and adds more romanticism to it.

Sales center design concept is "interstellar space", which use non-linear design elements, bold segmentation and surrounded by space, completely changed the original space. Is no longer boring, but to create a similar space, such as interstellar space, the space thus has a variety of ways, which also opens numerous possibilities.

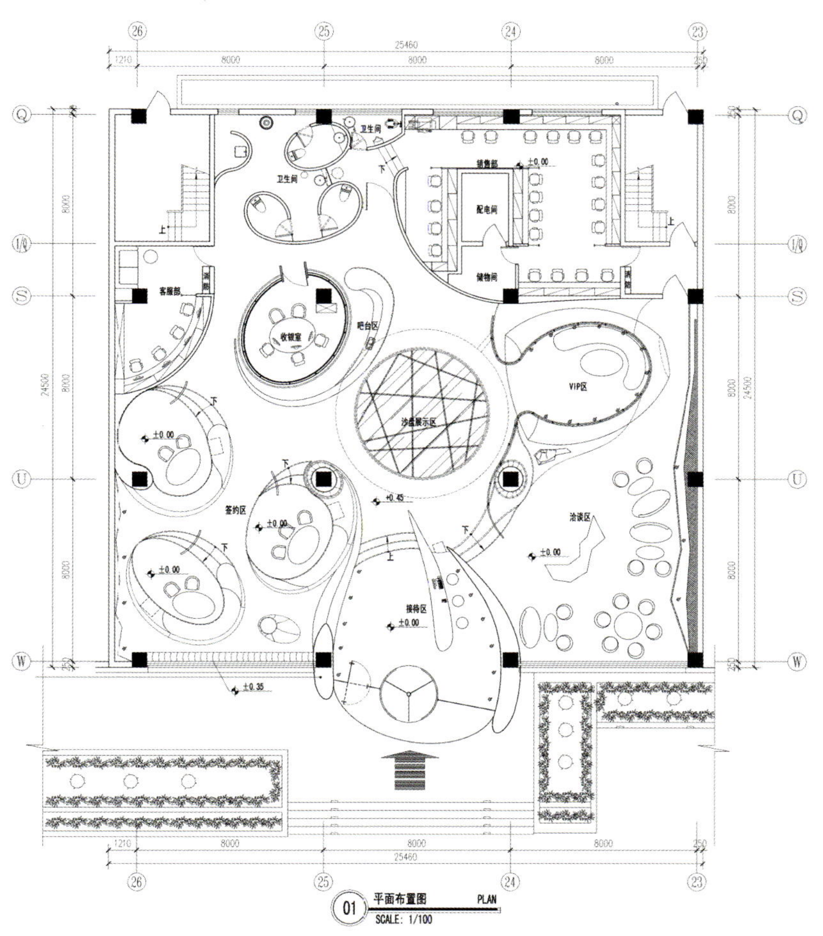
01 平面布置图 PLAN
SCALE: 1/100

A SALE CENTER
天墅销售中心

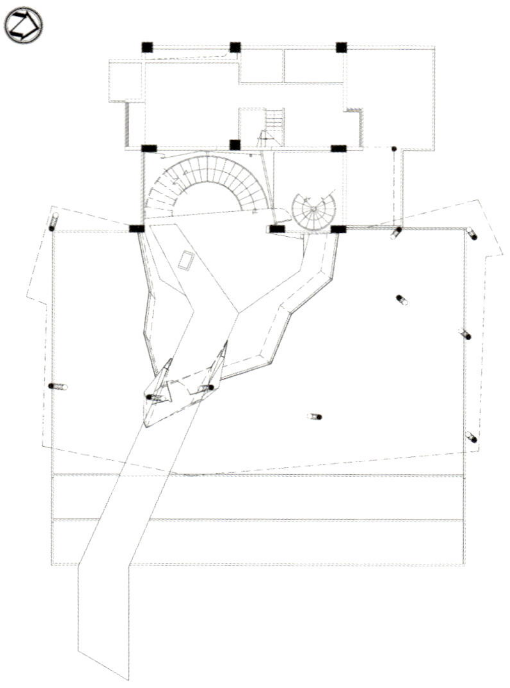

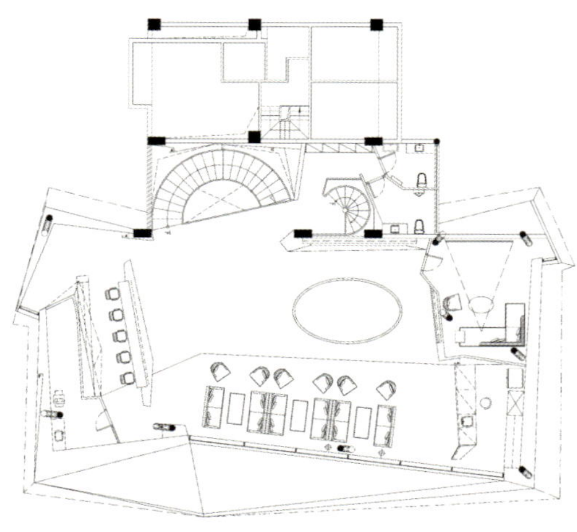

工程名称：天墅销售中心
坐落地点：厦门
面　　积：600 m²
设计公司：KLID达观国际设计事务所
主要材料：氟碳金属漆、橡木染色、玻璃、鹅卵石
竣工时间：2010
摄　　影：周跃东

天墅以空中别墅作为楼盘特点，为大众描画出市中心的新生活。而其售楼中心的整体设计理念也承袭了空中别墅的概念，并由此演化出“漂浮”的理念，创造了一个极具后现代风格的售楼空间，成为厦门又一处建筑景观。整个建筑体现了后现代建筑的美学观念，巧妙地运用“水”元素为建筑铺陈，最终通过建筑将水、阳光等自然元素与人相关联，形成一个极富艺术魅力的互动空间。

由于项目地点位于寸土寸金的厦门市中心，周围环境条件不佳，有旧公寓以及工地现场，正对面是一所学校，四周围并无景观可言。如何创造一个景观带来与建筑相匹配，也是设计师需要考量的。最终，设计师从建筑、室内、景观三方面整体构思，让三者相互延伸，相互作用。建筑为钢结构，以便更好地表达设计理念，力求建筑设计与室内设计手法的整体性与统一性。

The real estate located in Xiamen, is famous for its Villa Air which can draw a new life in the city center for the public. Overall design concept of the sales center also inherits the concept of the villa air, and thus evolved a "floating" concept, creating a very post-modern style for the sales space, which becomes another landscape architecture in Xiamen; the whole building reflects the aesthetic concepts of post-modern architecture. Designers cleverly use of "water" element as the conception of the building, and ultimately through the construction make water, sunlight and other natural elements associate with people to form a highly artistic interactive space.

The project is located in Xiamen which is high cost of land. How to create a landscape to match the buildings is what the designers consider. Final, designers consider three aspects including architecture, interior and landscape, making the three mutual extension of the interaction. Construction for the steel structure in order to better express the design concept, architectural design and interior design seeks to approach the integrity and unity.

Because of the "floating" concept, the sales center will be pulled to a high degree of layer 2, and the main function room is set up on the second floor. The building suspended in the air like an irregular box, changes shape and is full of sense of power. When deal with the landscape, designers make a large area of lawn together with a triple layer of progressive iterative pool, and pave the bottom with pebbles, making the sales center like floating in the water supported by the column. The entrance door is full of sense of geometry, designers use irregular shape to design the ceiling of the first layer, and with lights embedded in the ceiling. Using light as the space connection, designers weaken the color contrast naturally. Passing through the corridor, interviewers can go to the hall sales in the second floor by the rotation of the stairs.

When handling internal space of the second floor, irregular lines and graphics are still the main expression, while the contrast between color and material expression is the emphasis. Different colors, splicing overlap between irregular shapes make the main desk full of artistic atmosphere. Meanwhile, the surface is made of dark wood flooring and Light-colored tiles, in the conflict make the limited space extended. Different refraction, connection by using materials with perspective effect, and some large windows, make the construction has an excellent vision.

CROSSOVER ESTHETICS, A RECEPTION CENTER IN KAOHSIUNG

高雄都厅苑接待中心

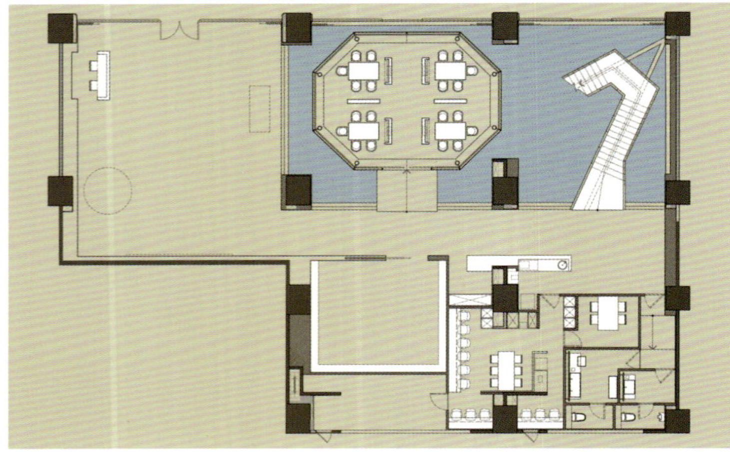

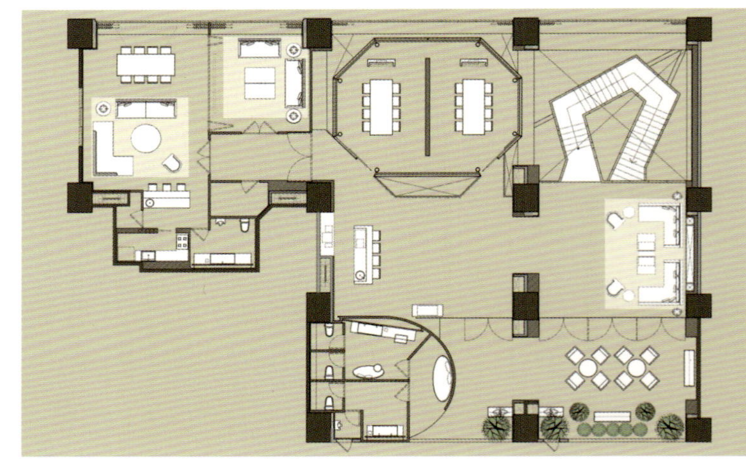

跨界（Crossover），一个时下具挑战性的创作概念与表现形式，当来自不同领域的元素，打破既有的框架并穿梭在各个层面，时尚vs汽车、艺文vs科技……往往激发出令人惊艳的创意。但若是两个本质上存在冲突与矛盾的元素，是否仍然可以作跨界的结合？本案以新的观察角度与思考逻辑切入接待中心的设计，先置入策划展览的主题，再导入空间规划与使用机能，企图在“艺术”展览与“商业”销售之间取得连结与平衡点，对应在设计面是以“类美术馆”的概念作为空间的布局，集结中国大陆、日本、韩国、菲律宾、印尼与超过半数的台湾当代艺术家，共21位30余件作品的“亚洲当代艺术新势力”作为展览的命题与空间内容，在三个单位、二个楼层的空间框架中，形塑空间、人与艺术品对话交流的氛围，引领空间使用者进入艺术欣赏、空间涵构与生活品味交织融合的场域，进而拉近大众与艺术品之间的距离，提升对台湾当代艺术家的认同。

门厅

门厅以无接缝的盘多磨地坪搭配纯白色的连续墙面作为空间背景，12 m^2的空间仅陈设三件作品：胡栋民回旋成圈的不锈钢雕塑《流水》，常陵气势恢宏的《山水大宏图》与李真饶富童趣的《无忧国土》。透过线条纯粹、尺度宽广的大厅空间，提供艺术品展现内在意涵与观赏者发挥无限想象的场所。

走道一侧的白色大理石咖啡吧台周边分别展示林建荣的摇头人偶《ZZZ-Ⅳ转头》、徐晓燕的油画《盛开》与牟柏岩的树脂雕塑《天空》。其中，设计者藉由端景墙上倾倒歪斜的墙体和天空的背景图像，进一步诠释《天空》挣扎向上攀爬的视觉效果，并透过镜面不锈钢的反射影像，将视角看不见的上方面部表情传达给观赏者，引起观赏者的会心一笑与共鸣。

玻璃立方体

7 m高、贯穿两个楼层的洽谈室是一由钢结构与强化玻璃组成的玻璃立方体，经由

黑色亮面抛光砖与镀钛不锈钢板构筑的镜面无边际水池的倒映与反射，不仅延伸室内空间的景深，并且营造出玻璃立方体悬浮水面的视觉效果。另外，为了呼应高雄作为海洋城市的城市精神，引入波光粼粼的海洋意象，玻璃立方体表面贴覆由设计者与林冠名一同为本案量身制作的影像作品《Sea》。强调私密性的空间特性搭配隐约可见的深海影像，成为同时兼顾使用机能与艺术表现的最佳组合。

白色的钢骨楼梯蜿蜒向上，连通接待中心上下楼层。设计者规划悬吊楼梯的结构钢缆作为投影屏幕，使林冠名的影片作品《泉》的影像与声音在钢缆之间流转穿梭，搭配楼梯本身的斜向支撑、不规则造型和胡栋民的不锈钢雕塑《迎面》，将楼梯转化为一座声光具备的巨型装置艺术。

公共整容间

公共整容间是以由设计者与吴季璁共同参与创作，为本案量身制作的作品《光之旋舞》为设计主轴。透过可更换投影主题的特制投影装置，在高雄市都市纹理转化而成的镜面分割图案上，投射出流转移动的影像。观赏者置身其中，同时感受时间、空间与影像的交迭重合，使静态的公共整容间呈现出多样的空间表情和不规则的律动感。

工程名称：都厅苑接待中心
坐落地点：高雄苓雅区
面　　积：1F 545 m^2，2F 530 m^2
参与设计：陈任远、陈敏媛
设计单位：动象国际室内装修有限公司
主要建材：盘多磨、钢刷染色木皮、镀钛不锈钢、钢琴烤漆、灰镜、大理石(卡拉拉白)、抛光砖、柚木板、地毯等
摄　　影：ABS color

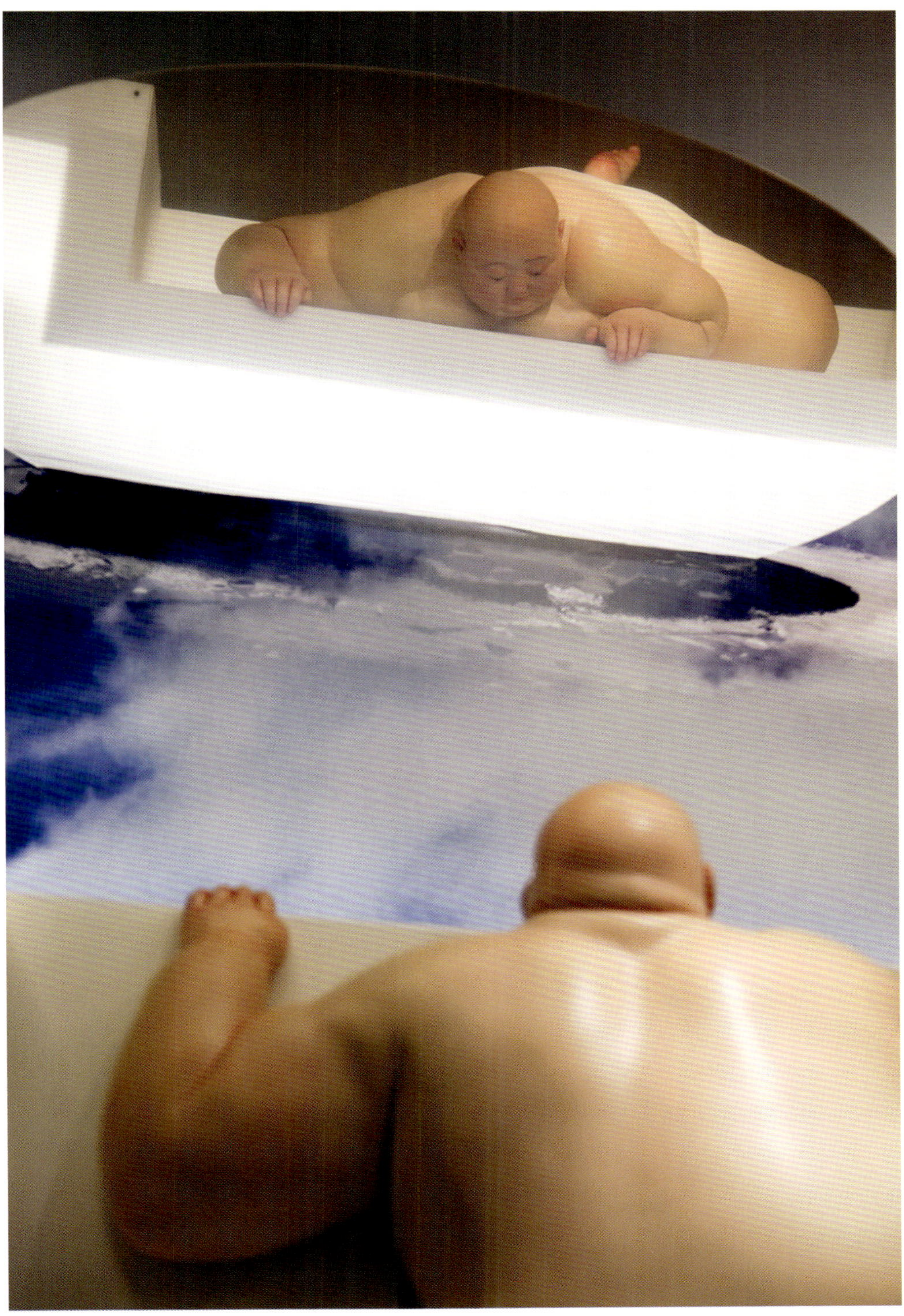

Crossover, the concept of a challenging and creative expression. When the elements from different fields, breaking the existing framework, such as: Fashion vs cars, art vs science and technology ... ... tend to produce stunning creative. However, if there are two essential elements of conflict and contradiction, whether as a combination of cross-border it? The case with a new perspective on the design of reception centers, first into the theme, and then import and use of spatial planning function in an attempt to "art" and "commercial" link and balance between, corresponding to the design surface is "Museum" The concept of a space layout and assembly in China, Japan, Korea, Philippines, Indonesia and Taiwan, the works of contemporary artists, form the "new forces of Contemporary

Art in Asia", in the space frame, the shape of space, people and art atmosphere of dialogue and exchange, lead the user into the artistic appreciation of space, space fabric and the integration of lifestyle interwoven field to narrow the distance between the public and works of art to enhance the works.

The hall floor with seamless walls with pure white background as space, 12 sqm furnished room only three pieces: Hu Dongmin stainless steel sculpture, "water", Changling's "great grand landscape"and Li Chen's "no Worry about land. "Through the lines of pure, wide-scale hall spaces for art to show the inner meaning of the place.

One side is the white marble corridor coffee bar, shaking his head around the dolls are displayed Lin Jianrong "ZZZ-IV

turned around, " Xu Xiaoyan of the oil painting "bloom"and Mu Bai rock sculpture "Sky. " Among them, the designer through the crooked walls and the sky background image, further interpretation of the "Sky " struggling to climb up the visual effects, and reflection through the mirror image of stainless steel, will not see the expression of perspective to convey to the viewer, causing The viewer's sympathy.

Glass cube

7 meters, through two floors of the negotiating room is a glass formed by the steel and glass cube, through the black shiny stainless steel polished and composed. This design not only extends the interior space of the depth of field, and creating a glass cube suspended water visual effect. In addition, to echo the spirit of the city, the introduction of the sparkling ocean images. Surface mount glass cube case review by the designer tailor made for the video work "Sea". Stressed the privacy of space characteristic of the

match, the looming image, as both functional and artistic expression using the best combination.

White staircase winding up, connectivity, reception center. Designers use the cable as a projection screen, so that works, "Spring " of video and audio in the cable between the shuttle, with its irregular shape of the stairs and stainless steel sculpture, "head", the staircase into a giant sound and light with Art Work.

Public cloakroom

Public cloakroom is held by the designer and the joint creation of Wu Chi, and to work, "Light Spinning Dance" for the design of the spindle. Can be replaced by the theme of a special projection equipment projection, the texture map divided into the mirror, moving images projected circulation. Watch who you were in the same time feel the time and space coincide with the images overlap, so that static space showing a variety of expressions and the law of irregular movement.

LUDI ZHENGZHOU STATION PLAZA SALES CENTER

绿地郑州站前广场售楼处

项目名称：绿地郑州站前广场售楼处
坐落位置：郑东新区高速铁路客运站西侧广场附近
业　　主：绿地集团中原事业部
竣工时间：2010.10
设计公司：MRT DESIGN
设 计 师：颜呈勋（Bill Yen）
摄 影 师：MOSEMAN ELEANOR ELIZABETH
面　　积：2000 m^2
主要材料：金属、大理石、丝质布料、地毯

项目位于郑东新区高速铁路客运站西侧广场附近，郑东新区商住物流园区内，北至七里河南路，南至广场北街，东至东风东路，西至心怡路。基地东侧为郑州新客站，北侧为七里河，西侧为未开发建设用地。站前广场项目共有9宗土地，此次获取的地块为D1、D2、D3、D4号地块。项目由写字楼、商业、酒店、公寓、SOHO等多业态构成面积达40万m^2的综合体项目，此项目属于郑州市政府重点关注的建设项目之一，未来将成为新郑州的中心地带。

售楼处的建筑外形由不规则的钻石切面组成，一块黄色的石头经过切割露出不同部位的玻璃切面，整个建筑充满立体的动感。

室内的设计也是从建筑的外形开始演变的，在设计的初始我们想象一块天然的水晶经过切割出现的各个立面，整个室内的平面按照功能规划后中间的大块区域留给模型展示和洽谈，所有的墙面都是经过切割形成的大块切面，傍晚灯光从建筑切面的各个部分溢出来，像TOYO ITO设计的TOD'S那样如同一个夜晚发光的宝盒。

天花和墙面我们采用大量的水晶切割纹理，天花在一层至三层大量的采用亮面的金属板，同时切割纹理在金属板上发生作用，使整个顶部如同许多的切割面叠加，看上去晶莹剔透；当金属的天花和柱体连接时，水晶切面被柱子拉扯下来，充满动感。切割纹理还充分的应用到卫生间的墙面、天花的发光灯具、VI系统等根据各种规能的需要发生不同的作用。

在材料的选择上采用高雅的色彩搭配；浅咖啡色的金属和白色的亮面大理石，以及大量的带有华丽感的丝质布料，通过这些材料以及设计逻辑，创造出一个现代而前卫的空间并不失丰富感和华丽氛围的商业空间。

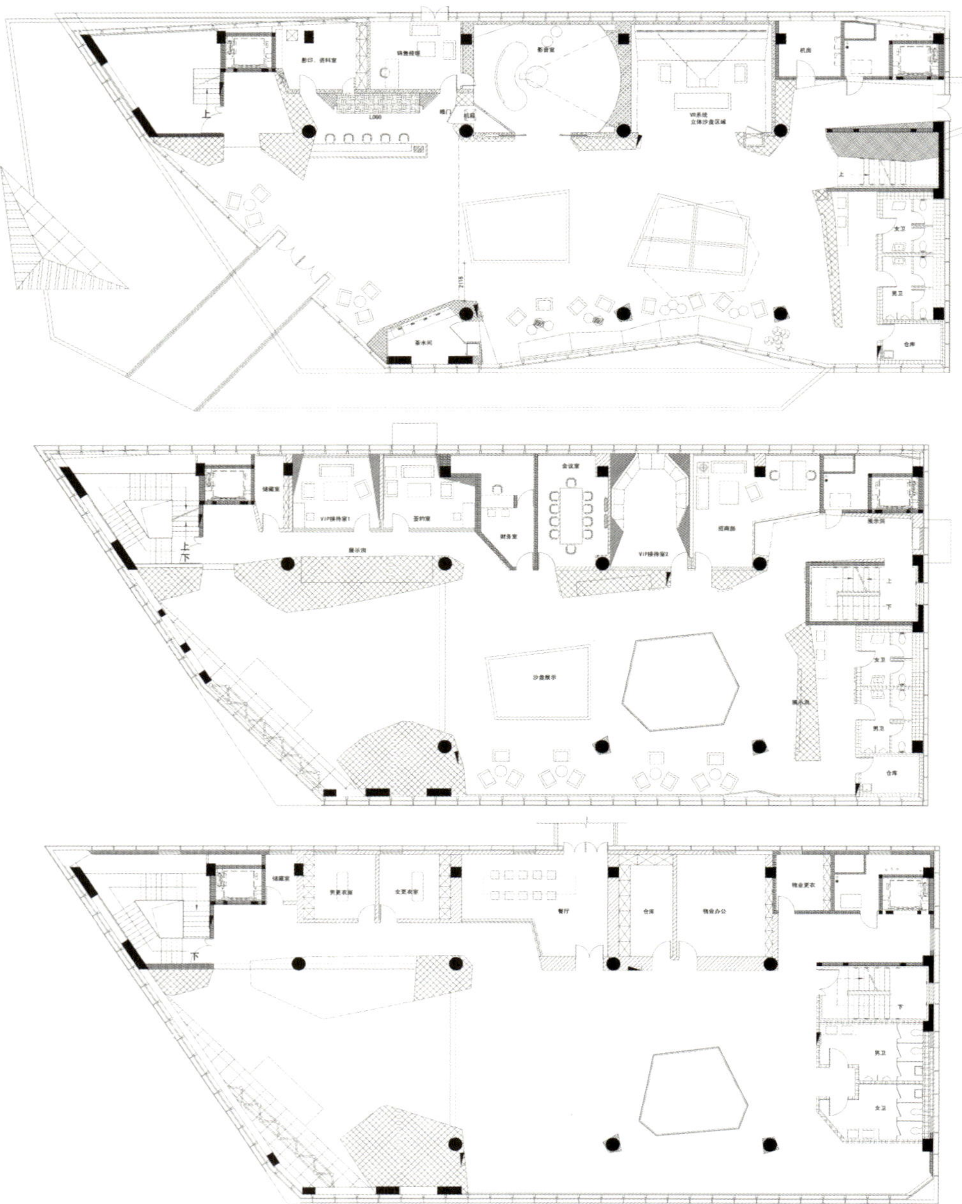

The project is located in High Speed Rail of Zhengzhou.The Square project is located in Zhen Dong New Area.It located in there,which north to Henan Road, south to Plaza north street, east to Dongfeng Road East, West to Xinyi Road.The north of it is the Seven Mile Creek,the west is the undeveloped land for construction. The project have nine land in total, we acquired for the D1, D2, D3, D4 plot for now. Project consists of office buildings, commercial, hotels, apartments, SOHO and other formats constitutes. The complex project has an area of 400,000 sqm. This project is focused by the construction of Zhengzhou city government on one of the future will become the new center of Zhengzhou.

Sales offices in the building foemed by the irregular shape of the diamond section.It is cutting to a piece of yellow rock exposed in different parts of the glass section, the whole building is full of three-dimensional movement.

Interior design began to evolve from the building shape.The initial of the design, we imagine to use a natural crystal to cut to different face.The entire interior of the plane in accordance with the function of planning to leave after the middle of the large area to display and discuss. All of the walls are formed by cutting large section.The evening light overflowed from various parts of buildings.It

just like the design of TOD'S TOYO ITO as the treasure box as a night light.

For the ceiling and wall textures We use a lot of crystal to cut.We use a large of shiny metal plate, while cutting the texture play a role in the metal plate, so that the whole top of the stack, like many of the cutting surface,It looks clear. Cutting is also fully applied to texture of the bathroom walls, ceiling light fixtures, VI system. It according to the needs of a variety of rules can place a different role.

The choice of materials used in color. light brown shiny metal and white marble, and a large number of silk fabric with a gorgeous sense, through the logic of these materials and design to create a modern and avant-garde space no loss of richness and gorgeous atmosphere of commercial space.

TRENDY INTERNATIONAL INTERIOR

大同玺苑接待中心

设计概念

“大同玺苑”位于南港经贸特区中，除人文景观及展现未来的价值性，在2010年后繁忙的生活圈中找寻人们所渴望的生活，7598 m^2植光计划打造出绿色奢华盛宴，本案的设计概念是人、环境、建筑之中寻找出一种平衡，彰显自然的视觉效果，时尚艺术品味的思维，以骑楼式建筑的概念，环绕着景观中庭的圆拱回廊，彷佛穿梭于大自然之中。透过空间、人与艺术品的互动，拉近来宾与艺术品之间的距离。结合艺术与生活，让空间显得更有生气，更具有人文素养。

空间说明

立面外观

接待中心外观以L型建筑外观及回廊造型包覆着未来的景观中庭，展现出一种保有健康与舒适的空间，让建筑物与自然环境全然融合。并以圆拱回廊的方式，营造于行进间满满豪宅气势与特区意象，来宾一走进此接待回廊立刻能感受到未来南港经贸特区中的绿色植光计划奢华盛宴。以白色作为主轴，纯净视觉及心灵，使来宾将外纷扰逐渐沉淀，让重心回归于生活层次，给予空间宁静的价值。成就简洁、时尚、自然，达到高度融合内部与外部空间的视觉延伸。

1F接待大厅

经由回廊步道，进入高6.7 m的接待大厅，空间内立有面贴木皮染色一气呵成的高耸墙面及天花，给人一种气势磅礴之感觉，但更展现出一种内敛，不饱和的原始天然，低调却蕴含能量，在充足的光照之下，彷佛树木般的呼吸着流动气息，与接待大厅毗连着的艺术廊道，挑高7.2 m的圆拱回廊及艺术造型墙面交错所呈现轻盈通透的空间感，户外阳光洒落在翠绿的草坪并搭配台湾海枣映照在白色艺术廊道上展现细腻、内敛，增添空间自然的彩度及视角的延续性，透过香杉实木长凳的摆设呈现出原始森林自然气息，搭配着装置艺术及艺术品的相较呼应之下，营造出品味、艺术、时尚、自然之气息。

1F销售及VIP区

本案采用个别式包厢设计，期能使来宾感受同等尊荣及礼遇并兼顾个别隐私及避免干扰，从自然的草坪沿生成为空间的发展性，藉由温润木质的触感，天然石材的细腻纹理，融合质朴洞石的肌理，呈现空间感的自然与舒缓，使来宾透过材质、接口、颜色，从现代繁忙的生活中找寻出自然的语汇。清透自然的雪花石灯罩搭配着沉稳低调的金属灯架，唤醒了自然空间中的时尚现代气息。

项目名称：大同玺苑接待中心
坐落地点：南港经贸特区
设计单位：动象国际室内装修有限公司
设 计 师：谭精忠
参与设计：洪国智、许玉臻
面　　积：外观面积 4311 m²
1F面积 2451 m²
2F面积 548 m²
主要材料：石头漆，细石漆，洞石，皮革，钢刷染色木皮，金属漆，镀钛不锈钢，灰镜，大理石(雪花银狐、黑网，深棕灰，黑檀木)，人造雪花石，抛光砖，木地板，地毯

洽谈室以悬浮弧型天花、木皮圆柱、景观大窗，链接着户外的绿景，营造出轻盈、自然的视觉效果，塑造身、心、灵的调和，让人与人之间可自在的对话，柔化忙碌的生活情绪，使空间释放出透明、释压的气围，藉由内外空间的对应，产生空间与人的感觉互动因子，而使空间艺术化，形构出视觉的延伸性。悬吊天花上清透自然的雪花石，就像点缀夜晚时天空上的星星。

VIP室透过壁面的木纹、质朴的洞石、原木的地坪、通透的景观大窗交错的呈现户外的绿景，摆脱了生冷的原样，喧吵着人也是大自然的一环，延续着优雅从容的步调，阳光和煦的迤洒入室，使来宾沉浸于悠闲缓慢的人文情调勾勒自然不做作的迷人风情。

玻璃电梯及大梯

伴随着自动门的开启，白色墙体，矗立在翠绿的草坪，透过户外的光影绕映在玻璃大梯上，彷佛绿景忽出于眼帘之间，随着踏上悬浮阶梯，清楚鸟瞰着户外景色，白色天花营造出一种步步云端之气息，旋回般的楼梯，有种耐人寻味的感觉。

两侧的白墙呼应之下，使得空间有种延伸的味道，玻璃电梯在挑高10 m的建筑体中，将呈现户外的阳光绿景的融合及夜晚的光影变化。

2F艺术廊区及户外休闲区

在通往样品屋时，将踏入空间、材质与艺术品的对应下，展现出一种在艺术馆中低调沉稳的人文气息，橡木、皮革与布料更是勾勒出休闲自在的氛围，衬托出沉淀后的繁杂心灵，回归于自然谐和。一眼遥望景观中庭及圆拱回廊与四周高耸的树木交集之下，彷佛进入了世外桃源，阳光缓缓在洒落在地板上，享受难得的自然芬芳及远璃都市烦嚣，点缀出生活、环境、人文息息相关。来宾可细细品嚼的“家”的感觉。

整容间

深色的大理石肌理，些许的金属框架，搭配着木皮质感，交谱出时尚都会男性的豪迈气息中带着返普归真的纯真，更与白色天花，清澈的明镜与沉稳的灰镜双重呼应下，展现出空间的时尚、自然、纯洁。彷佛在烦乱的都会中找寻到一种解放。

纯洁雪白的大理，朴质的洞石，搭配着木皮质感，再与通透明亮的雪花石下，表现出现代女性的冰雪聪明，浑然天成的搭配下，塑造出现代城市中可找寻到的一种身、心、灵完全解放的空间。

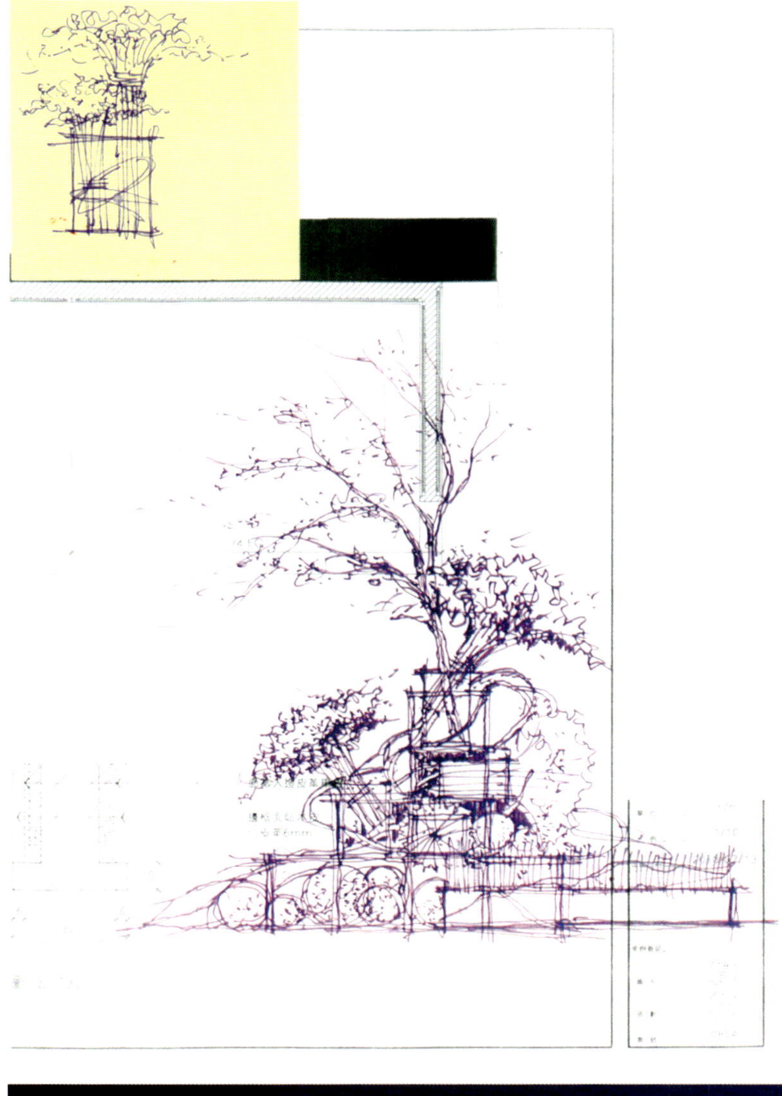

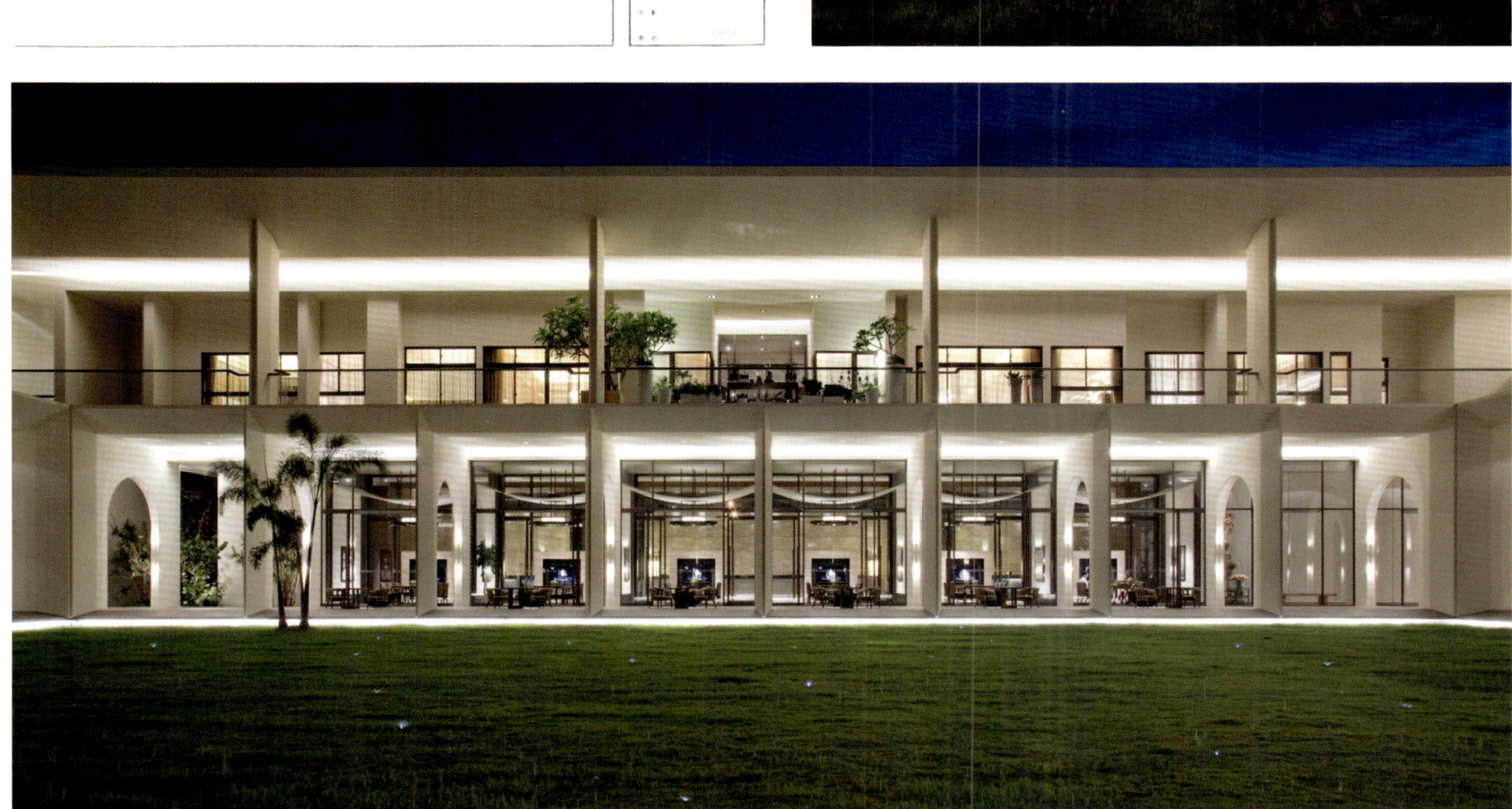

It located in special economic zone in the South, in addition to cultural landscape and to show the value of the future, after the peak in the living area of 2010, people is finding the desire of life, 7598 sqm to create a green floor luxury feast, the design concept of the case is to find a kind of balance between the people,environment and building, To highlight the nature of the visual effects, fashion, artistic taste of thinking, the concept of arcade-style architecture, the landscape around the atrium of the arched corridors, as if the shuttle being in nature. Through space, the interaction factor between man and art, to narrow the distance between the guest and the works of art. Combination of the art and life, so that space appears more vibrant and humanities.

Facade Appearance: the appearance of the reception center and the building exterior corridors L-shape the future of the landscape covered atrium, showing a space to maintain health and comfort, so that the building completely integrated with the natural environment.when the guests entered the reception corridor They will mmediately feel luxury feast of green plants of the future economic and trade zone in the South Light plans.

1F reception hall: via corridor trails, 6.7 meters high into the reception hall, the space stands a surface veneer , I t gives us a feeling of tremendous momentum, but more to show a restrained, not saturation of the original natural. The incense cedar wood benches through the display showing the natural flavor of the

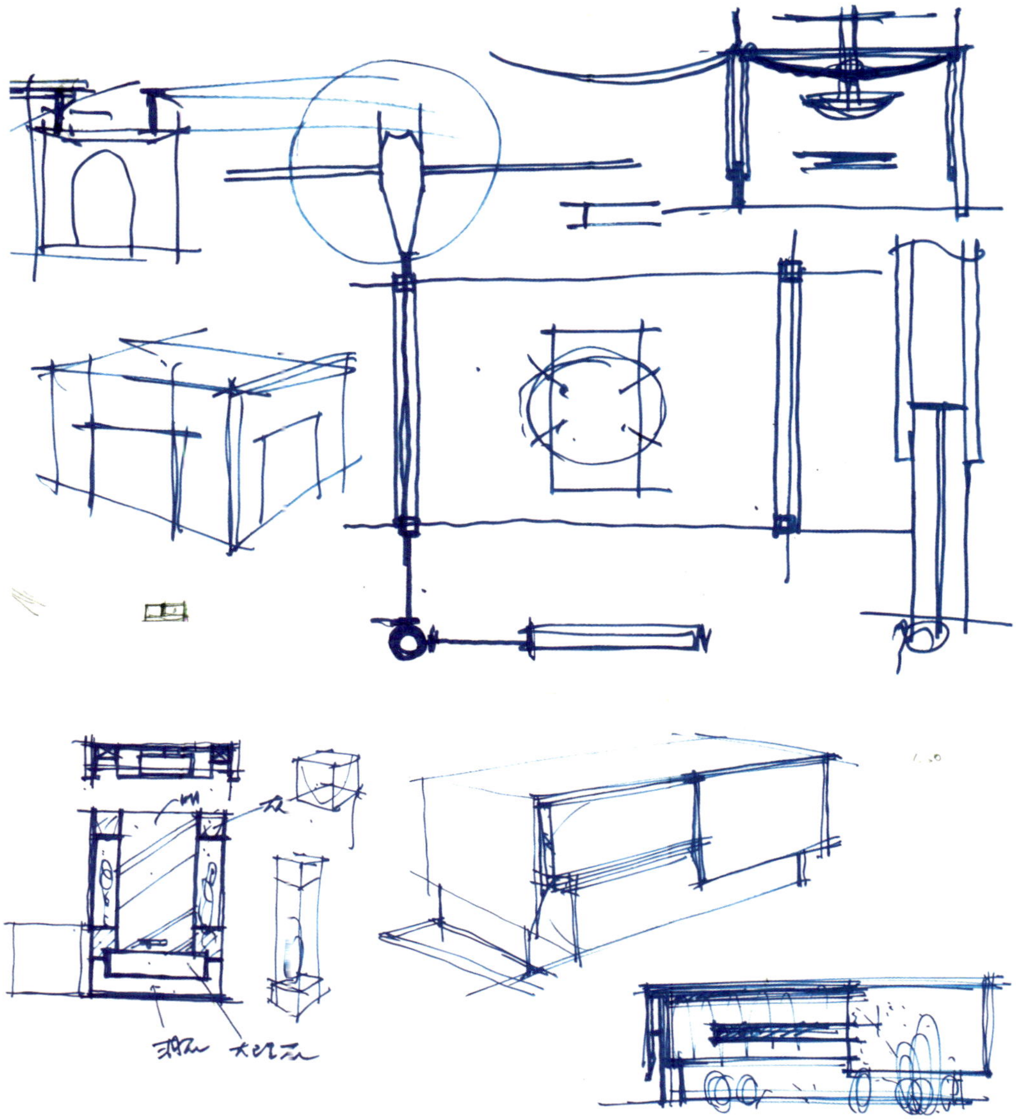

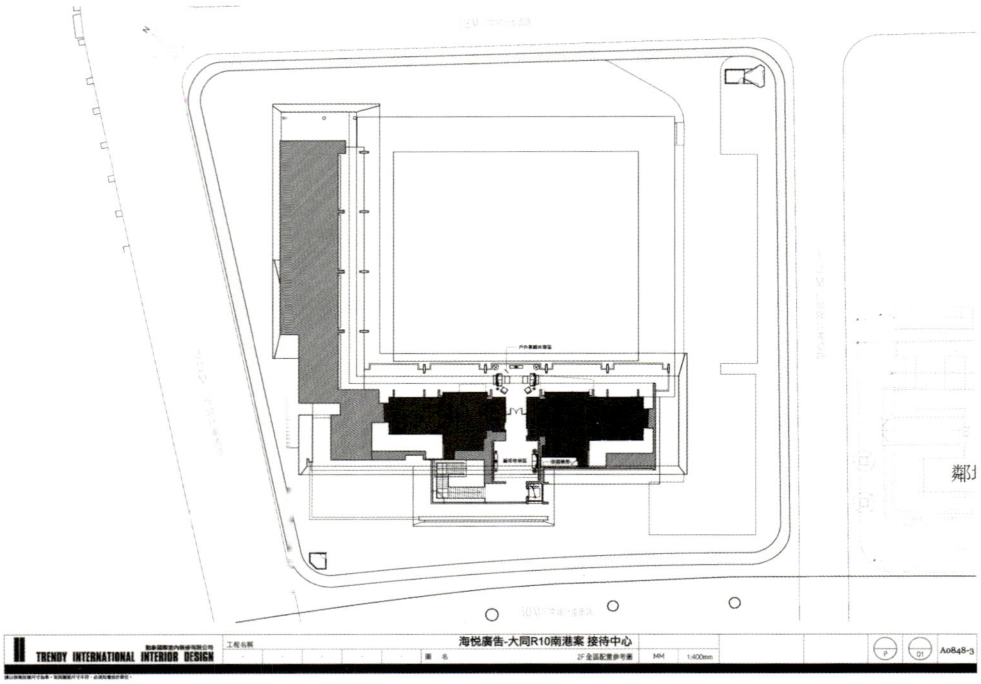

TRENDY INTERNATIONAL INTERIOR DESIGN

工程名稱 海悅廣告-大同R10南港案 接待中心

圖 名 2F 全區配置參考圖 MM 1:400mm

A0848-3

original forest , with the installation art and works of art under echo contrast, creating a taste, art, fashion, natural flavor.

1F sales and VIP area: the case with individual-type box design,To make the guests feel the same courtesy and consideration of individual privacy and avoiding interference. from the natural grass along the development of students into space, through the gentle touch of wood, the delicate texture of natural stone, travertine rustic texture fusion, showing a natural sense of space and ease, so that guests through the material, interface, color, from the modern busy life to find out the nature of the vocabulary. The regotiated room use the curved suspended ceiling, wood columns, landscape large windows, linked to the outdoor Greenview, creating a light, natural visual effects, It shaped the body, mind, and sp rit of the reconciliation, so that the dialogue between people can be comfortable , soften busy life emotions, to make room for the release of a transparent, release pressure of the air around, by the corresponding internal and external space, resulting in the feeling of space and human interaction factor, leaving space for art, and constitutes an extension of the visual.

VIP room through the wall of the wood, rustic travertine, logs, flooring, large windows staggered transparent presentation of the landscape outside of Greenview, get rid of a cold in the original, noisy arguing who is also a part of nature, The pace continues the elegant style, sprinkled with sunshine of winding burglary, so the guests slowly immersed in the natural and unforced charm.

Glass elevator and a large ladder: With the opening of automatic doors, white walls, standing in green grass, outdoor lighting through the glass around the large staircase on the map, as if sud-

denly out of sight between Greenview, with the step on the suspended ladder, clear aerial view of the outdoor scenery, the white cloud ceiling to create a step by step, the atmosphere, cycle-like stairs, feeling a kind of interesting. Under the white walls on both sides of echoes, making the space a kind of extension of the taste, the glass elevator, in the building of 10 meters high, will present the integration of outdoor sunlight,greenview and night lighting changes.

2F art gallery district and outdoor recreation areas: access to the sample house, it will enter the space, the corresponding material and artwork, It shows a feelig like you are at the Museum of Art in a low-key cultural atmosphere of calm, oak, leather and fabric more is outlined casual comfortable atmosphere, bring out the complex after precipitation of heart and return to the natural harmony. An atrium overlooking the landscape and the arched corridors and the intersection under the towering trees all around, as if to enter the paradise, the sun slowly floating down on the floor.To enjoy the natural and peaceful life,It let us know the life, environment and culture is closely related. And also guests can get a feeling of "home".

Make-up room: a dark marble texture, a little metal frame, with a veneer texture, cross-spectrum of a stylish atmosphere in the city with heroic men, back to Cape Guizhen innocence, but also with white ceilings, clear and calm gray mirror mirror dual echo. It shows a feeling of fashion, natural and pure in this space. It is like to search for the upset to a liberation. Pure white marble, primitive stone hole, with a veneer texture, and then with transparent under bright alabaster. It is showing the modern women's smart. It shapes to look for a body, mind, and spirit to complete feel liberation of the space in the modern city.

TRENDY INTERIOR DESIGN

市政厅接待中心

精銳·市政廳

设计概念

“品牌”是现代企业经营环节中不可或缺的一环，其影响力涵盖企业精神、发展策略、经营管理、产品定位等层面。一个成功品牌的无形价值甚至能够超越企业有形的资产，并且可以存在更长远的时间。

本案的设计概念是引入艺术展览作为接待中心的空间接口，藉由对当代艺术的关注与赞助，使企业品牌与艺术文化产生链接。同时，透过当代艺术作品突破创新、引领时代的风格特质，接待中心不仅扮演提供房地产销售所需使用空间与机能的角色，也能提升企业品牌的价值与销售产品的形象，进而引发消费者的感性共鸣与消费动机。

以“动漫艺术”与“亚洲当代艺术”为双主题的展览，网罗台湾、中国、日本、韩国与印度尼西亚的当代艺术家，共19位20余件作品。包含科学小飞侠、无敌铁金刚等动漫主题的作品，唤起20世纪50's ~70's人们的共同记忆，而席德进、朱德群等当代大师的作品则是吸引专业收藏家的注目。空间、人与艺术品对话交流的氛围，不仅提供观赏者共同的话题，拉近大众与艺术品之间的距离，间接也提升消费者对企业品牌与销售产品的认同与好感度。

空间说明

1. 立面外观

接待中心外观以覆贴白色塑铝板的ㄇ形框架为主体，边框与楼板的斜面造型减轻40 cm墙板造成的厚重感，地面上黑色亮面抛光砖与镀钛不锈钢板构筑的无边际水池的倒映与反射，则营造接待中心悬浮水面的视觉效果。夹纱玻璃围塑而成的玻璃立方体由框架中间悬挑而出，在夜间熠熠发光宛如悬挂天际的珠宝盒。

2. 1F门厅

1F入口门厅是由镀钛不锈钢、清玻璃与烧面鲸灰石所组成的透明玻璃屋，灰白色调搭配简洁的线条，环绕在无边际水池与水面的雕塑艺术品中间，散发出空灵清澈的气息。过道的空间性质，仅有接待柜台与悬吊雕塑艺术品，后方H型钢加灰玻璃的钢骨楼梯提示通往配置在二楼的主要空间的动线。

3. 2F大厅

挑高11 m、放射状的造型天花板，展现楼梯空间的气势。2F大厅以皮革与镀钛不锈钢边框装饰墙面，办公室、洽谈室等空间以隐藏门隐身墙后，保留连续的墙面以展示包括成太镇、曾庆熙、尹钟锡与天野喜孝的科学小飞侠在内的动漫主题画作。低调奢华的壁面材料围塑完整不受干扰的展览场域，展现人文、品味、尊贵的空间深度与质感。

项目名称：市政厅接待中心
坐落地点：台中
设计单位：动象国际室内装修有限公司
设 计 师：谭精忠
参与设计：陈任远、陈敏媛
面　　积：1F 面积 1270 m^2
2F 面积 960 m^2
主要建材：铝塑板，皮革，钢刷染色木皮，镀钛不锈钢，灰镜，大理石(意大利白)，抛光砖，柚木板，地毯

"Brand" is an indispensable part of modern business. Its influence covers entrepreneurship, strategy, management, product positioning and others. A successful brand can even go beyond the intangible value of the tangible assets of enterprises, and there can be a longer time.

The design concept is the introduction of the case art exhibition space as a reception center.To conbine the brand and the arts and cultural to be together. Meanwhile, a breakthrough innovation through contemporary arts.To lead the styles of the era, the reception center to provide real estate sales not only play use of space and function required for the role, but also can improve the value of corporate brand image and sales of products.It is also can approve people to get emotional resonance and consumer motivation.

Facade Appearance: the appearance of reception centers to stick to the mo-shaped white plastic sheet as the main framework, border and floor of the ramp shape to reduce 40cm thick wall due to a sense of shiny black tiles on the ground with the titanium plate constructed of stainless steel No margin reflecting pool and reflection, the reception center to create the visual effect of suspended water. Yarn made of glass, plastic clip around the middle of cantilevered glass cube out of the frame, gleaming in the night sky like a hanging jewelry box.

1F hall: 1F entrance hall is a titanium stainless steel, clear glass and burning face whales transparent glass house composed of limestone, white tone with a simple line around the pool and the water in the absence of the marginal middle of the sculpture works of art, send out the clear ethereal atmosphere. The nature of the aisle space, only the reception counter and hanging sculptures, plus the rear

H beam steel gray glass staircase leading to the configuration prompts the main space on the second floor circulation.

2F Lobby: with a ceiling of 11 meters, the shape of radial ceiling, stairs to show momentum space. Hall 2F titanium stainless steel frame with leather and decorated walls, offices, rooms, space for negotiations to hide the door stealth wall, retaining wall straight into the Pacific to show.It including the town of Zengqing Xi, Yin Zhongxi and Tian ye xi xiao.It including animation theme paintings. Low-key luxury plastic wall material around the exhibition field a full uninterrupted, to show culture, taste, distinguished spatial depth and texture.

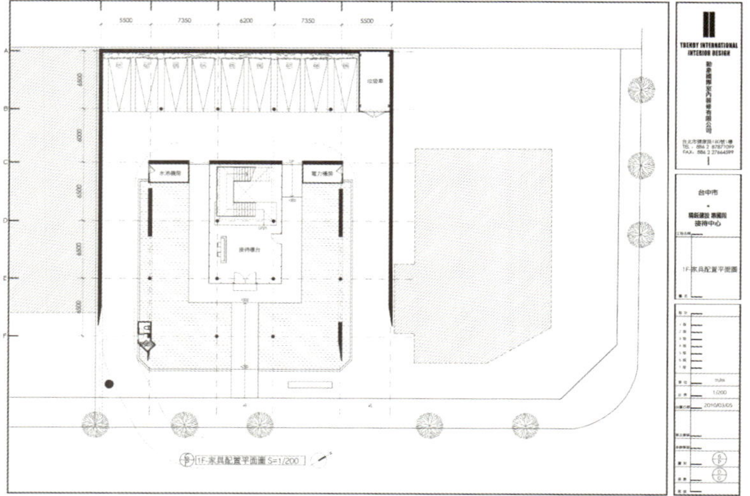

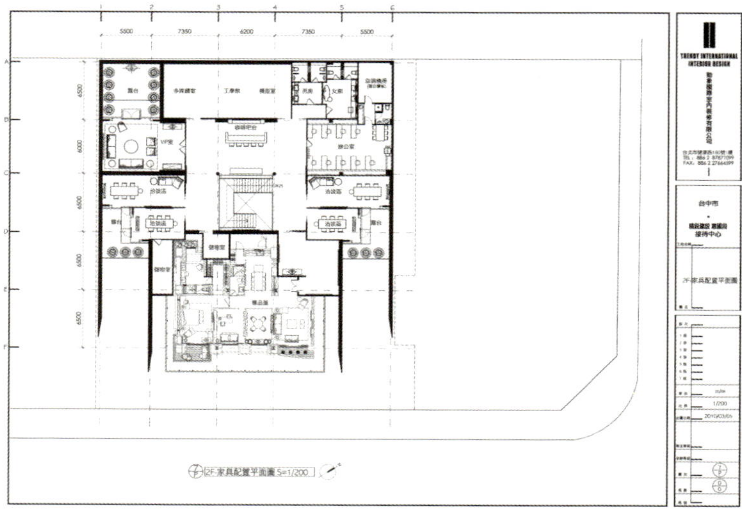

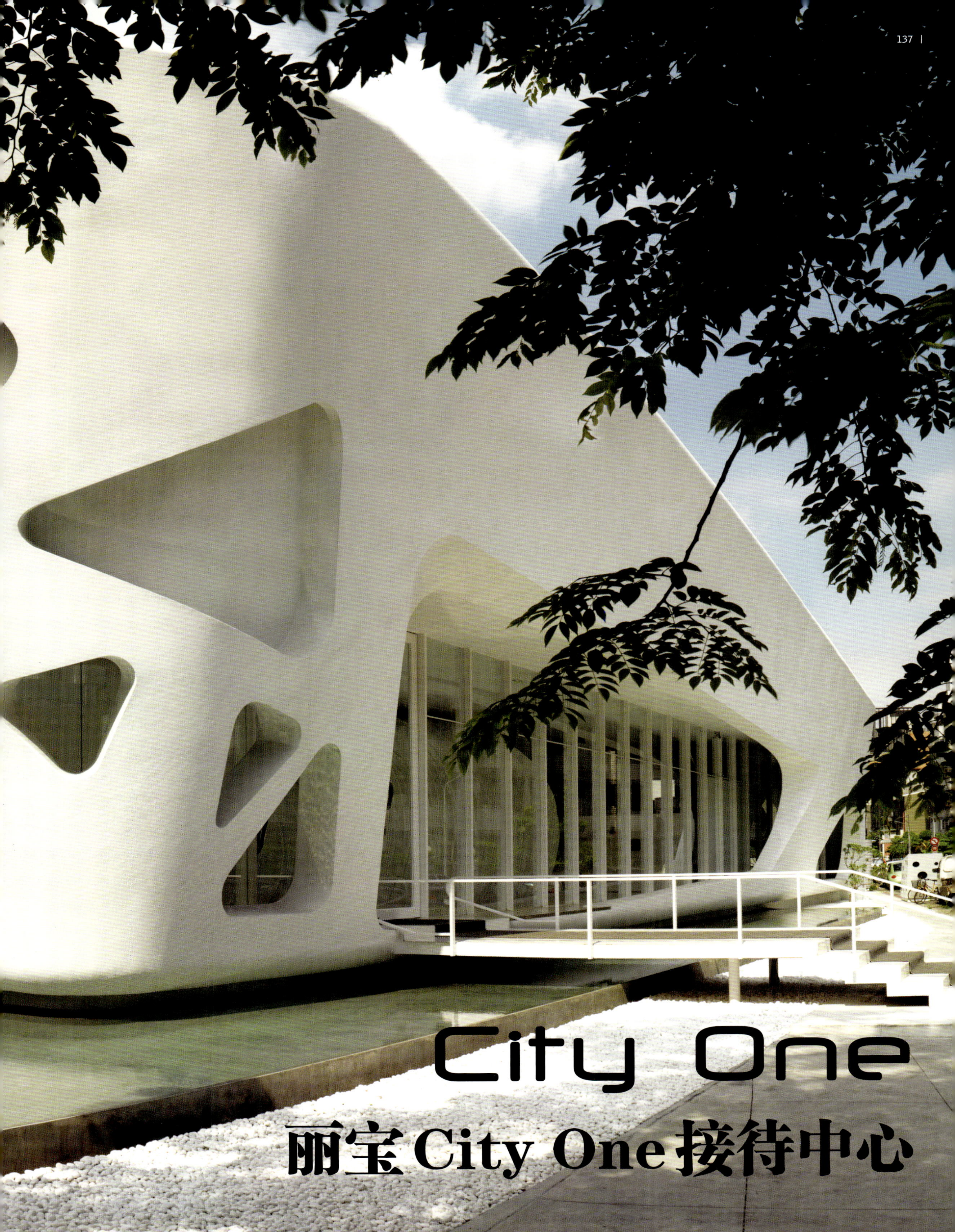

City One

丽宝City One接待中心

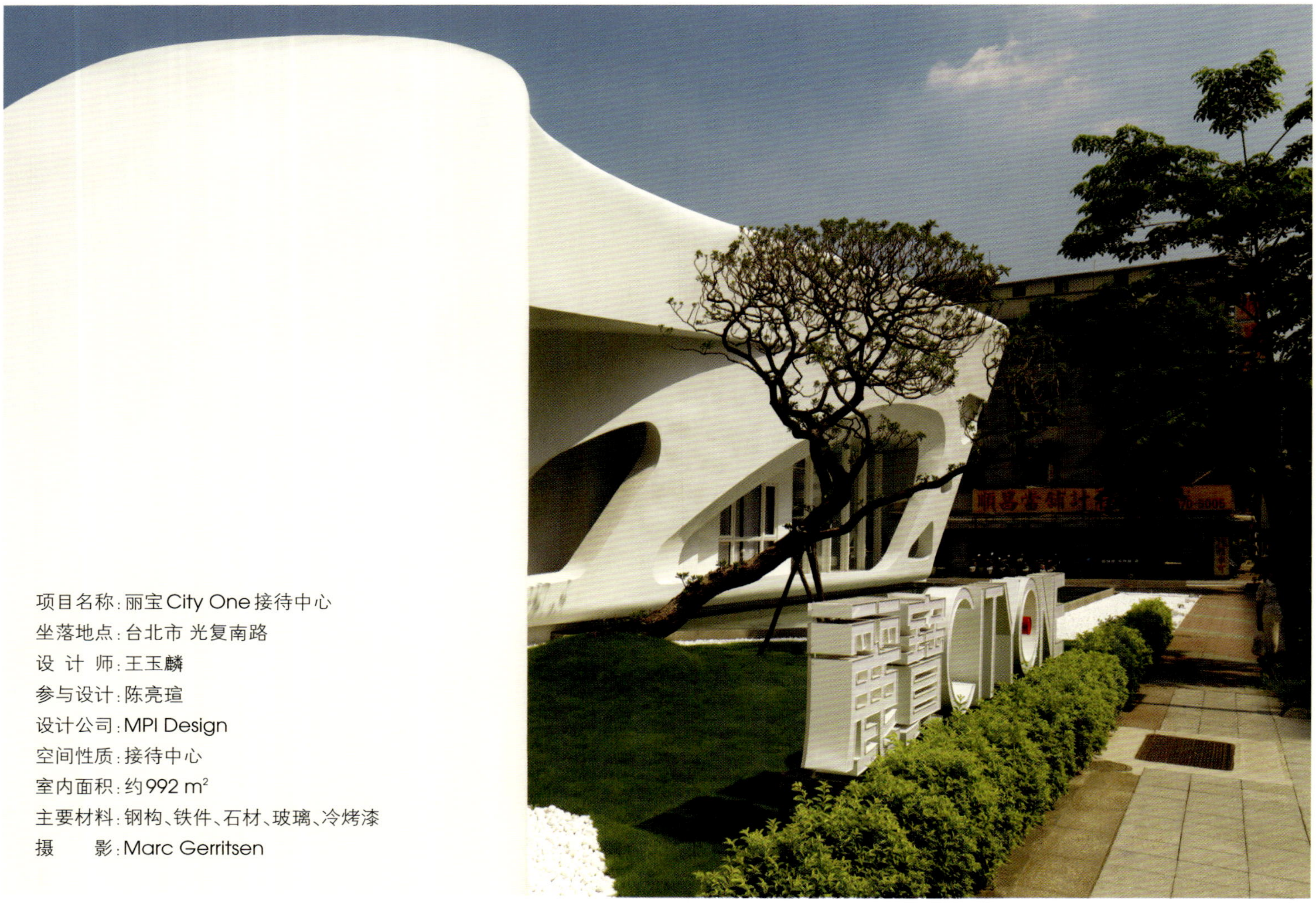

项目名称：丽宝City One接待中心
坐落地点：台北市 光复南路
设 计 师：王玉麟
参与设计：陈亮瑄
设计公司：MPI Design
空间性质：接待中心
室内面积：约992 m²
主要材料：钢构、铁件、石材、玻璃、冷烤漆
摄　　影：Marc Gerritsen

这是一个住宅建案的销售中心，地点位于台北市东区光复南路及市面大道高架桥边，一个两面临 路一面临6 m巷的基地，对住宅建筑来讲，条件并不理想，基于这样的利地条件下，改造环境及视觉是设计师需要思考的主要课题(吸引众人的目光)，这件事有很多种手法。

蝴蝶穿着彩衣五颜六色，炫丽夺目，这是利用色彩达到目地的手法。街上的车辆众多，哪一部能引起视觉的目光，除了颜色，速度及造型是主要的因素。

我们选择了速度及造型，若与蝴蝶的绚丽做比较，我觉得较有深度。

动感的造型当然是这个案例的设计主轴，在三面临路的条件下，纯白色的量体及动感的曲面 造型相当耀眼。这栋临时建筑于刚完工下架时，经常引起路口交通问题，这个路口因此有了常态的交通警察站岗，也多次因为路人不经允许私自拍照留念而与当下的警卫产生纷争。某种程度来说，它已经达到设计的主要目的，当然房屋也在很短的时间销售一空。

This is the case of a residential building sales center. It located at Guangfu South Road, Taipei, and the Eastern Avenue viaduct side of the market. It is facing a two-faced 6-meter Lane Road. A base of residential construction. The conditions are not ideal.Bur based on this conditions. The transformation of the environment and visual need designers to think that is the main topic "(to attract people's attention) There are many ways".

The butterflies wear a colorful coat. They use color to catch the ideal. There are many cars in the street, and which car catch the eyes? That is the main factors, it is color, speed and style.

We choose the speed and shape, it's brilliant, if compared with the butterflies, I think it will be more depth.

Dynamic modeling course to become the key idea of the design in this case.Facing the three choices. The pure white body and the dynamic surface modeling quite dazzling. When the newly completed of this temporary buildings on the shelf. It is often causing the traffic problems in the junction. So there is a normal traffic police standing guard in there.It also have disputes with the current guards because many pedestrians take pictures without permission. To some extent, it has reached the main purpose of the design, of course, the house is also sold out in a very short period of time.

TRENDY INTERNATIONAL INTERIOR

国泰天母接待中心

项目名称：国泰天母接待中心
坐落地点：台北
投资兴建：国泰建设
销售企划：国泰建设
设计单位：动象国际室内装修有限公司
设 计 师：谭精忠
参与设计：陈任远、陈敏媛、张耀伦
面　　积：1F 面积 850 m^2
2F 面积 500 m^2
主要建材：仿石漆，夹纱玻璃，铁管烤漆，染色木皮，雪花石，大理石(意大利白)，抛光砖，铁木，地毯，灰镜

设计概念

"Location, Location, Location!"是房地产销售的名言，从古到今、由东方到西方皆能适用。对销售者而言，它是高价热销的保证，对投资者而言，它是选择标地的不败法则，但是对居住者来说，却是生活其中的真实体验，它代表的是生活机能、空间情境与历史情感的复合体。

天母，台北传统的豪宅聚落，临接中山北路七段的一千余坪的方正基地更是无可取代。但是，在优越的硬件环境条件背后，天母的生活该要如何的想象？

因此本案的设计概念是企图藉由接待中心与样品屋空间的演译，提供天母生活的想象；艺术家的艺术创作描绘，唤回天母人文的蕴涵以及摄影师的影像呈现，重新定义天母历史的精神。

空间说明

1. 立面外观

接待中心外观经由铁板烤漆收边之线性折板，以由外而内、渐次分层退缩的方式，勾勒出两侧停车场与入口悬浮玻璃盒的空间轮廓。夹纱玻璃围塑着由框架中间悬挑

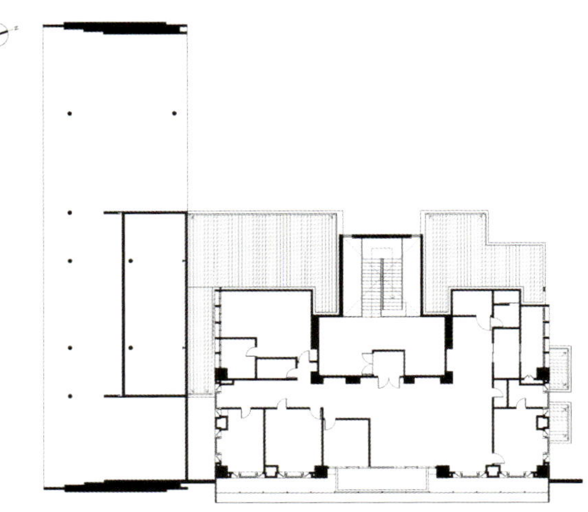

而出入口悬浮玻璃盒，在夜间宛如熠熠发光的珠宝盒。地坪则以同色系石英砖由内而外延续铺设至楼梯踏阶，营造接待中心一气呵成的空间延伸视觉效果。

2. 大厅

1F入口门厅以刻意放大的空间尺度，简洁利落的线条，搭配两座摆设艺术品与景观盆栽的四柱亭，与三幅由画家为本案量身创作，以“富春山居图”为灵感、天母周边山景为主题的“富春天母图”，充分反映本建案基地的大气度与大格局。藉由预先展示未来公共设施空间的艺术品“国泰天母图”、留设相同座向位置与大小的景观庭园、规划相同的车行动线，使参观者在预售阶段便能与之后完工的实际空间产生链接。

3. 廊道

连接各使用空间的廊道，以雪花石灯柱柱列营造低调典雅的空间氛围，两侧墙面则以摄影展的方式，展出由摄影师以[天母印象]为主题所创作的摄影作品，藉由摄影镜头的呈现，捕捉拥有丰富历史、异国风情和人文涵养的天母的过去与现在，让想要进住或已经住在天母的参观者都能够重新认识、发现天母的美。

"Location, Location, Location!" is very important to real estate sales since ancient times, from east to west. Taking into account the seller, it ensures high price, for investors, it is easy for them to choose a good estates. But for residents, however, it is one of the real life experiences. It is a hybrid, concluding function of life, space and historical emotion.

Tianmu is irreplaceable, a traditional luxury village in Taipei. However, behind the excellent condition of the physical environment, how can we imagine the real life of the Tianmu? Therefore, the design concept of this case is to imagine the life in Tianmu by reception center and sample house space. By watching the exhibition hold by photographers, art created by the artists, designers tell visitors what is Tianmu and re-define the spirit of history Tianmu.

Exterior: the appearance of reception center is decorated by baking painted iron plate, and it forms the outline of the suspending glass box located in the parking-lot on both sides and the entrance from outside and holds back every layer. The whole glass box which is hung in the mid-point is surrounded by a wired glass-made frame; and it has suspending entrance and exit, as if a jewel box shining in the night.

Hall: 1F deliberately enlarged the spatial scale of the entrance hall, using simple and neat lines, with two decorative plants, four-poster kiosks, and three sites created by the artist tailored for the case. Designers use [Fuchun Mountains] as inspiration, using mountains around Tian-

mu as theme, which fully reflects the construction's impressive appearance.

Corridor: The corridor connecting the use of space uses Alabaster lamp post to create a low-key and elegant space atmosphere. Two sides of the wall display the photography created by the photographers using "Tianmu impression" as the theme. With the presentation of cameras, by catching its rich history, exotic and culture, visitors know well about the past and present of Tianmu and people who want to live or have been lived in Tianmu are able to re-understanding of the beauty of Tianmu.

約地標

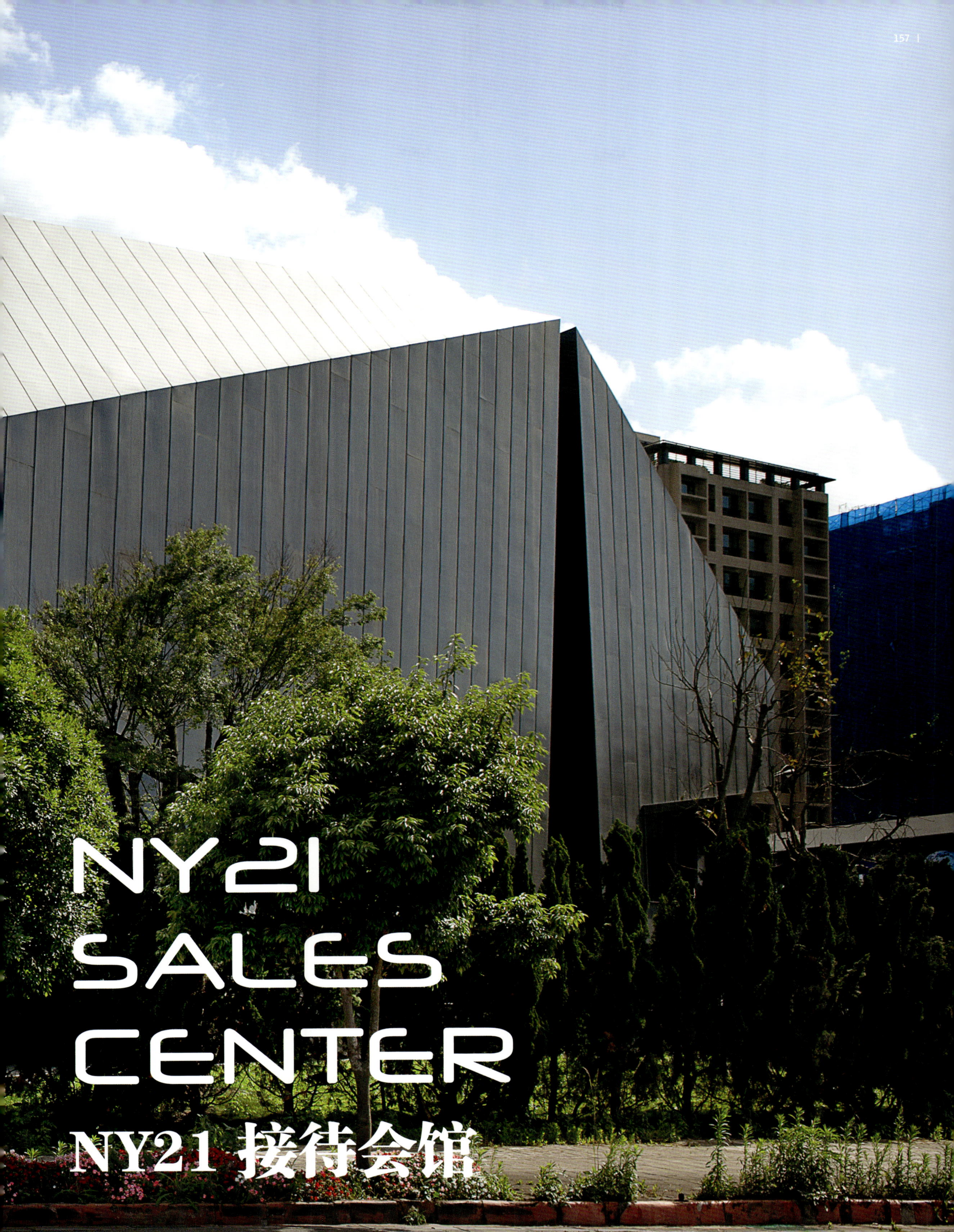

NY21 SALES CENTER

NY21 接待会馆

NY21 接待会馆位于道路交会的节点，呼应如此具地标性的位置我们希望创造出如艺术品般的雕刻体，使其成为一个醒目的装置艺术存在城市中。

接待会馆将主空间移至二楼，一楼则为回车动线，方便客户行进、停泊，在此同时也将整栋建筑物与地面拉开，让外观形成一颗半悬浮的雕刻体。连接三层楼面的楼梯贯穿了整个接待会馆也与造型连贯，形塑成量体切面结构、串联里外空间，使楼梯及墙面像绕着雕刻体在内部垂直延伸一般。开阔明亮的挑空区域，视觉感受由水平垂直向上，天光从顶流泄而下，形成明暗、虚实的空间架构。

项目名称：NY21 接待会馆
坐落地点：台湾台北市南港
设计单位：大尺设计工程股份有限公司
设 计 师：郭旭原、黄惠美
参与设计：黄敬哲
业　　主：海悦广告
空间性质：接待中心
建筑面积：约 752.43 m^2
基地面积：2265.82 m^2
主要材料：塑铝板，墨色玻璃，石材，氧化美板
摄　　影：Marc Gerritsen

NY21 reception hall located in the road intersection node, matching such a landmark location, we hope to create a sculpture-like construction, making it a striking presence in cities as a work of art.

Reception Hall moves the main room to the second floor, considering the convenience of customers, such as parking, moving, they designed generatrix for motor vehicle on the first floor. At the same time the entire building will be off the ground, so the appearance of the building looks like a body of semi-suspension. Staircase which connects three floors runs through the entire reception hall, connecting outside and inside, making the stairs and walls look like around the sculpture. The space area is bright and open, visual perception from the horizontal to the vertical, sunshine shines directly to the floor, all of them make the space structure with virtual and fact, Light and dark.

HYL GALLERY

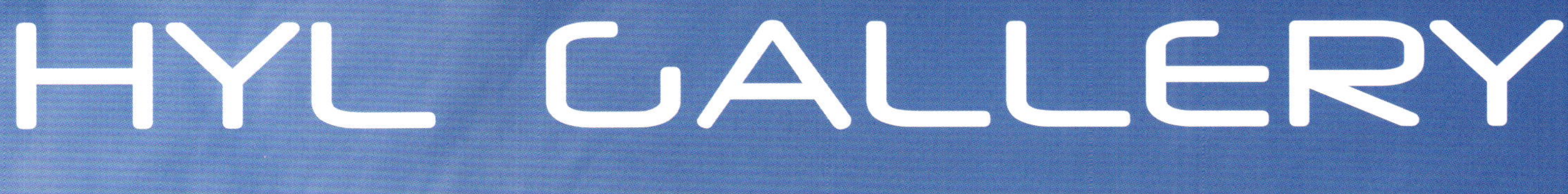

玖住章HYL Gallery 自主建筑创志馆

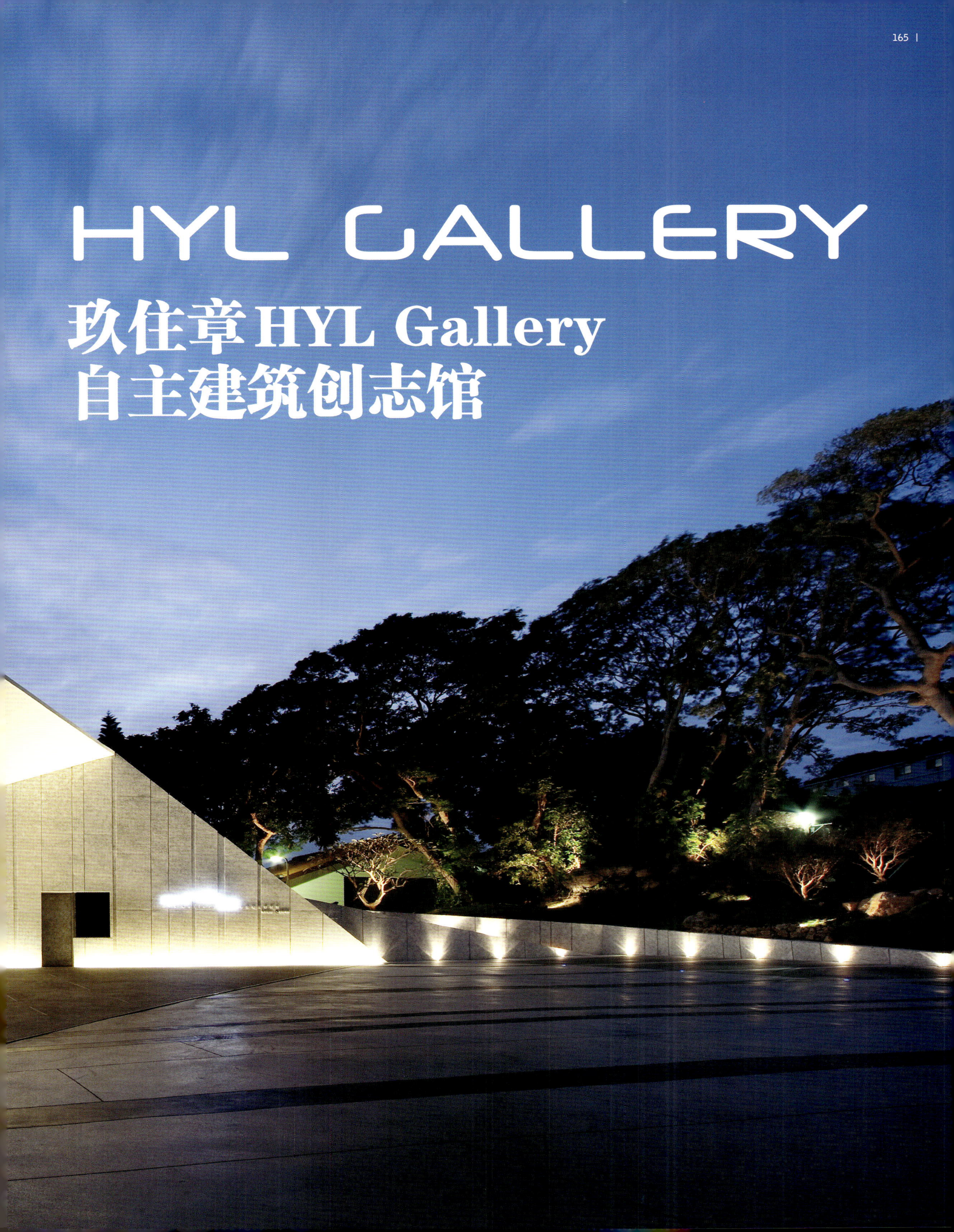

设计说明：

HYL gallery 由一系列不同比例之V-Shape式空间，自外形至内里，再从地面至高层，或压缩或伸张，或转折或攀升，形塑组织而成。在作为一可提供不同角度及高度之观景装置（Viewing Device）的同时，进一步突破传统售楼处的形式，以私人美术馆之姿，将三位建筑师的空间故事展演，顺势链接至基地大无限与无价的极致景观。

HYL gallery is formed by a series of different ratios of V-Shape style space. From the surface to the inside, then from the ground to the top, compress or stretch, turn or climb, and then make it. As a viewing device, HYL gallery can provide different angles and heights to watch the landscape, what is more, it breaks the traditional form of sales offices, as the name of a private museum, develop the space story of the three architects, linked to the priceless acme landscape.

项目名称：HYL gallery
坐落地点：台北市
设计单位：齐物设计
设 计 师：甘泰来
参与设计：陈海伦、张耀文
摄　　影：卢震宇
空间性质：售楼处
主要材料：石材、钻泥板、木皮、灰镜、epoxy
面　　积：1F 570 m^2、2F 32 m^2，共602 m^2
设计时间：2009.05~07月
竣工时间：2009.10

YUE TAI FENG FAN
岳泰峰范

项目名称：岳泰峰苑
坐落地点：台北 景美
设 计 师：王玉麟
参与设计：陈亮瑄
设计公司：MPI Design
空间性质：接待中心
室内面积：1404 m^2
主要材料：钢构、石材、玻璃、木皮、铁件、硅酸钙板
摄　　影：Marc Gerritsen

罗斯福路是一条历史悠久的文化大道，台湾大学也位于这条大道上。本案位处于这个历史悠久人文荟萃的文教区域，也因为这样的一个历史包袱此区已经呈现老旧的现象。这是我们设计思考一个很重要的因素，因为这个建案的量体及占地面积有一定的尺度，我们希望透过特别拉高达到3层楼高的空间比例这样的设计手法能将未来的实际环境及空间尺寸，完整的呈现在消费者面前，也藉由这样一个封闭式的设计手法能够排除历史视线的包袱，围塑一个内庭的空间，与环境做一个明显的区隔，透过刻意营造的云型开口将自然的光线与天气变化引进室内。让人虽然身处室内，还能和自然有亲密的接触，体会到自然无穷尽的变化，而四周墙面开的人型的造型开孔，除了能让带给空间一些幽默的趣味，增加空间的穿透感，并将未来小区的人文动态作一些呈现。

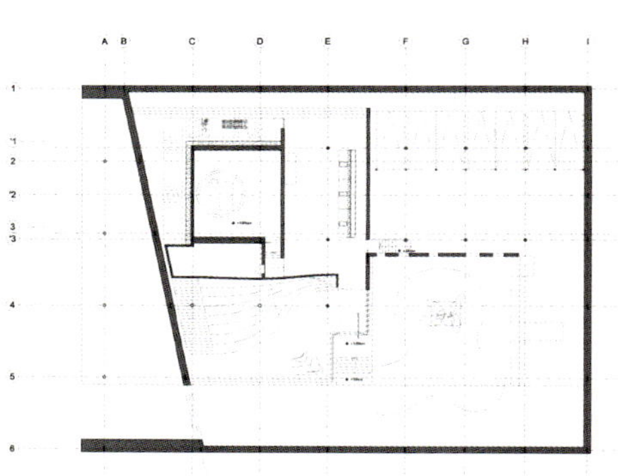

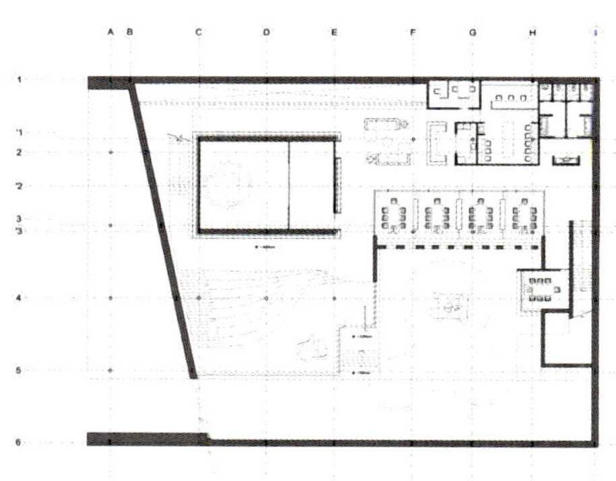

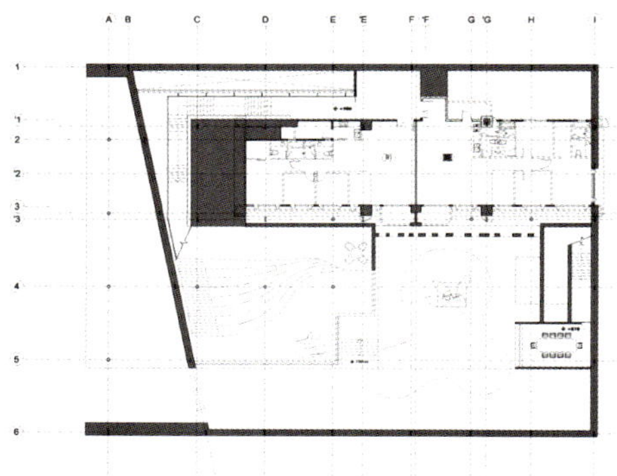

Roosevelt Road is a long history of cultural Boulevard, National Taiwan University, also located on this road.

In this case,it located in the historic, cultural and educational areas.Is also because such a historical burden of this area has been showing old phenomenon. This is a very important factor of our design.Because the case was built a certain amount of body and covers an area of the scale. We hope that through special space of 3-storey design techniques such proportion of the future can physical and spatial dimensions.The complete show in front of consumers, The design of such a closed approach to eliminate the burden of the historical line of sight.The environment as a distinct segment, through the deliberate cloud-type openings will be created by natural light.Although people living in interior, but people could feel they are closed to the natual.

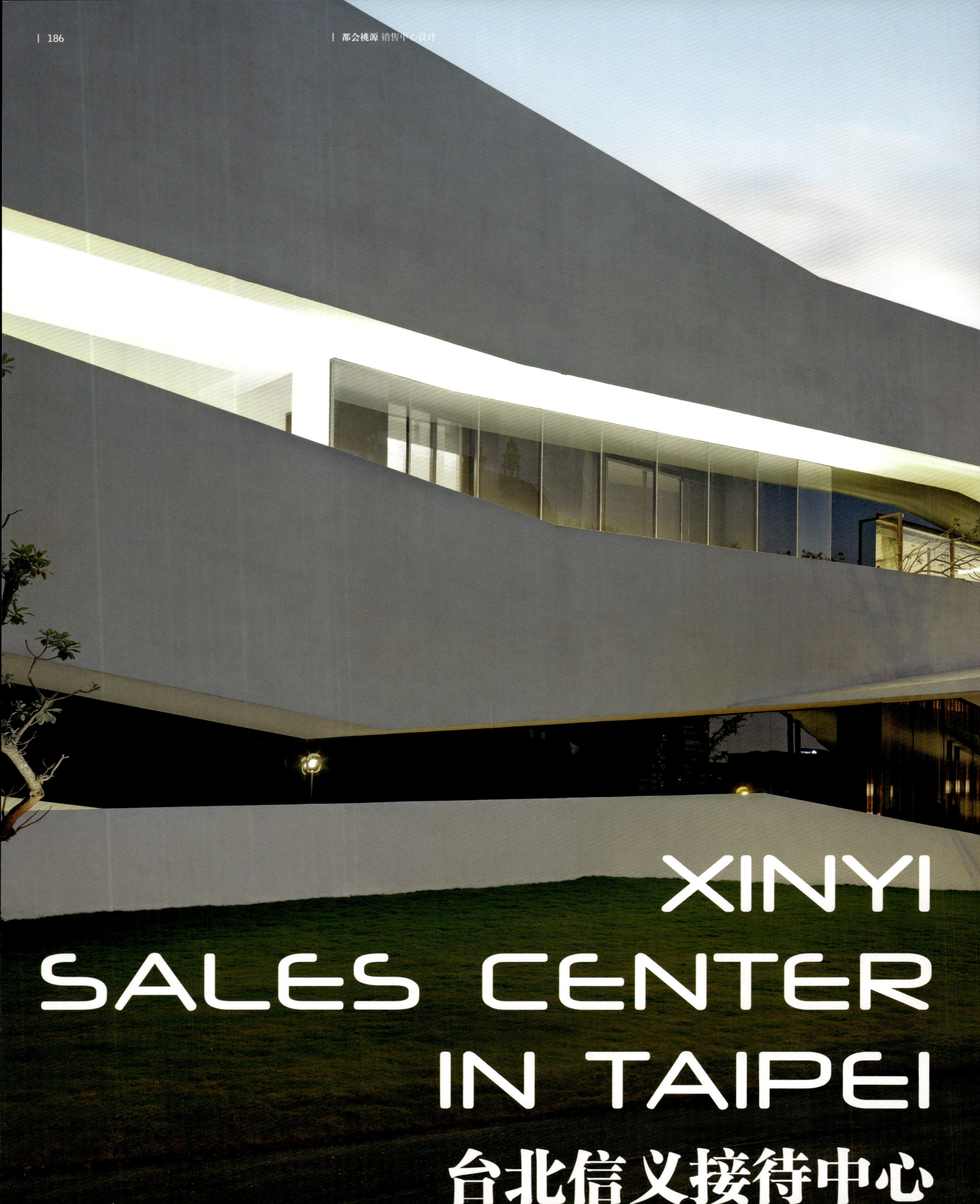

XINYI SALES CENTER IN TAIPEI

台北信义接待中心

"台北信义"的设计理念

-"形与意"-以山水园林为师

设计初始希望以这个案例来探讨长久以来思考中「形」与「意」的空间议题，形体的操作与意念的表达是建筑思考的手段与传递空间的命题，而东、西方文化的差异造成美学与空间有着极大的不同与发展。

西方建筑的思考逻辑着重于形与物质的探究，讲求辩证与科学的推理。从素描石膏像开始，光线、比例、配比、技法的研究，都是西方以方法论对美学所做的了解与认知，进而开始达到对空间环境的控制；而东方的空间观则常由人本身的意念与想象出发，追求物质环境以外的满足，探究空间能达到的境界。以山水画为例，中国画不在乎透视法与拟真性而寄情于山水画中的意境，在山水画中"可观"、"可居"、"可游"，画中空间有了无限的想象。此外东方文化亦深受禅宗的影响，画面中的留白，建筑中的"院"都留下极大的"虚"与"空"，在这虚与空之间，"人"的想象可以与千古的时间与空间相联系。

东西方文明与建筑空间的表达有着很大的差异，而在近代东方文明的崛起后，当代建筑更产生极大的冲击与变化。这次的设计藉用西方建筑中几何形体的逻辑认知与操作，和东方山水画中的空间意念融合，期望以抽象的山水园林作为设计的主轴。

在中国园林中，廊道与墙一方面连接各空间，同时阻隔过渡的元素。像浮云一般飘在天空中若即若离的廊道，高低起伏，转换人行径的视角，也改变活动的情绪。外观的形体飘浮流动，轻盈如行云流水，意图颠覆量体在型态上的重量感。

设计初始就希望高低起伏的景观能由外而内延伸，创造一个地景与建筑形态交融的空间。由外而内缓升的坡道带动视觉的流动，贵宾室、柜台、吧台都在形体上化为抽象的山水，而水由外部不经意的流进室内，在墨镜的交错运用下，产生梦幻般的效果。空间的端景与视觉的高潮是一座黑色往上的楼梯悬在水面上，而由上方洒下的天光让人在这里与自然又再一次的相会。

The initial design of this case want to explore the thinking which has long been "Form" and "meaning" of the space issues, The physical operation and the expression of ideas is the construction and delivery of space to think the proposition means, while the different cultures of eastern and western has caused a great difference and development in aesthetics and the space.

Logical thinking of Western architecture focusing on shape and material of the inquiry, and stress the dialectical and scientific reasoning. From the beginning the plaster sketch, light, proportion, ratio, research techniques, are in the West to understand the methodology of aesthetic and cognitive made, and then began to control the space environment; and Eastern concept of space is often made up of people starting their own ideas and imagination, the pursuit of the physical environment to meet other than to explore the realm of space to achieve. To landscape painting, for example, Chinese painting and realistic perspective of care and always focussed on landscape painting in the mood, the landscape of "significant", "livable", "can we go", the painting space with unlimited imagination. Besides Eastern culture also by the influence of Zen, the screen blank in the construction of the "Court" have left a great "virtual" and "empty" space in between the virtual and the "people" with the imagination time and space through the ages linked.

There has a ruthless big difference between eastern and western civilizations and the expression of architectural space. And the rise of oriental civilization in modern times, the contemporary architecture also have a great impact and changes. The design of use Western logic of geometry construction and operation of cognition, and landscape in the Eastern idea of integration of space, hoping to abstract the design of the spindle land-

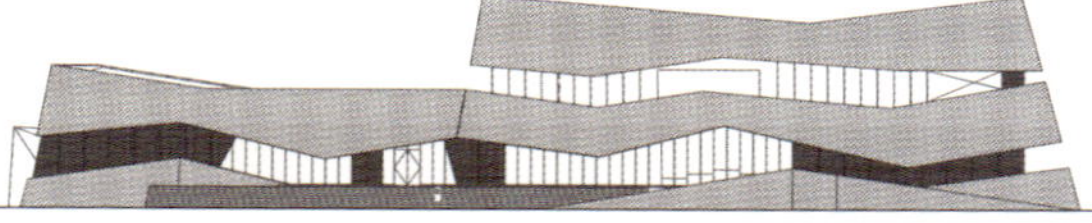

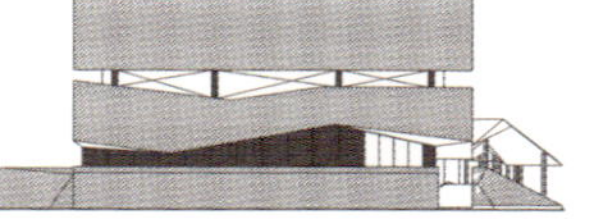

scape garden.

In the Chinese garden, the corridor connecting the space with the wall on the one hand, while blocking over the elements. Like a cloud floating in the sky generally ambiguous corridors, ups and downs, conversion of people acts of view, but also change the activity of the mood. Physical appearance of floating body movement, light and graceful, intent to subvert the amount of body weight on the type of flu.

The initial design of the undulating landscape want to extend from outside to inside, which creating a blend of landscape and architectural style of the space. Ramp from the ramp outside to inside flow driven vision, VIP room, counters, bar-type body in the landscape into abstract, and the water flow from the outside into the interior inadvertently, in sunglasses staggered operation \ using the following produce fantastic results. Landscape and visual side of space is the climax of a black hanging up the stairs on the water, and sky from the top of the shed where people once again came face to face with nature.

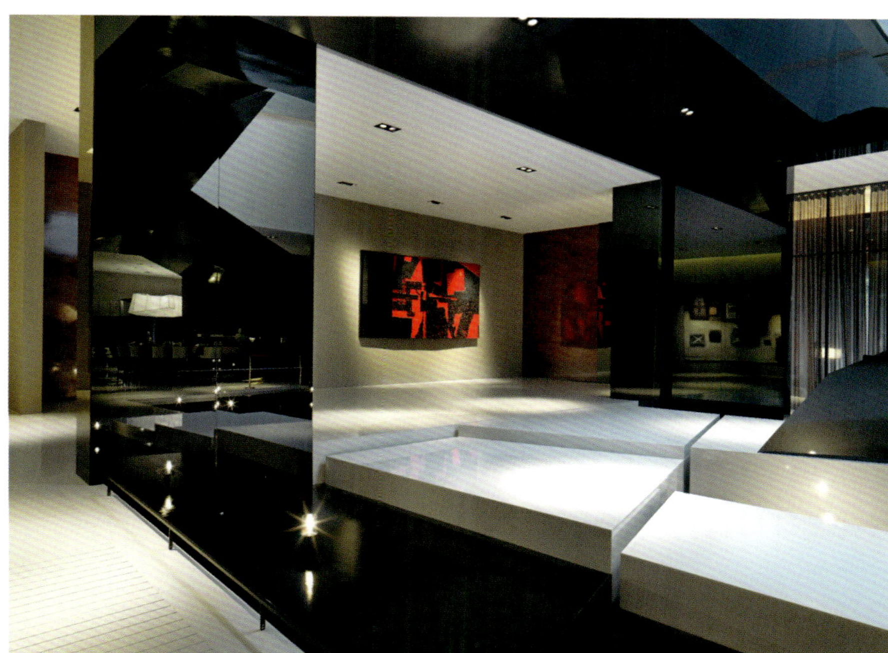

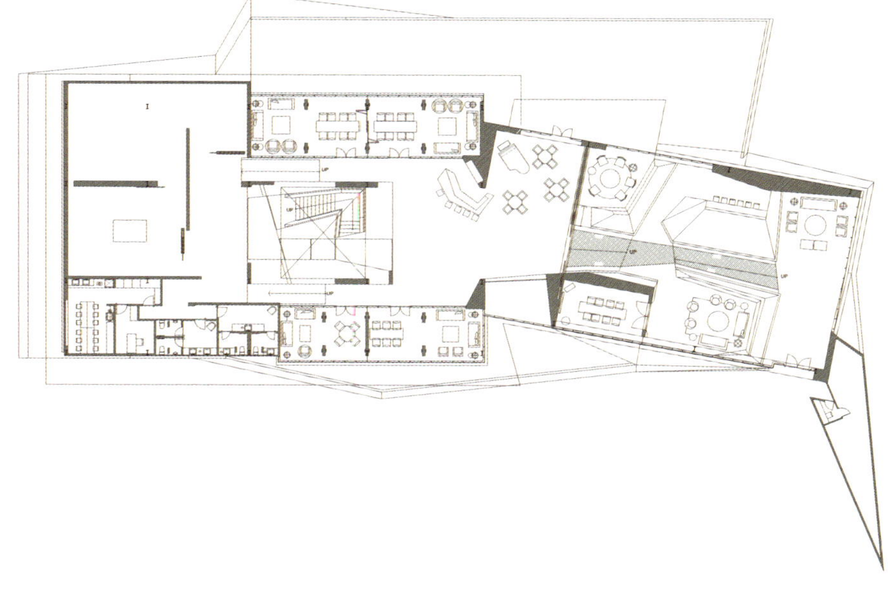

JIU YI SHI MEI SALES CENTER

九昱十美接待中心

本案位于台北市人文荟萃、自然山色富蕴的文山区，且基地也正好坐落在一具有自然坡面的环境中，故接待中心的设计上，自然将居所韵味及场域精神，定位于与文山区人文质感及自然环境相呼应。

进入基地的道路，依原始地貌，顺势开出斜坡，行进间，映入眼帘的即是一悬挑在基地上，横向展开的建物，呈现宽广面向，藉以衬托出基地之广阔；纯白之外观全栋正立面，以纯粹之格栅条柱之语汇 连贯，不同的视角，辉映不同的景色，试图将周遭自然景色，全揽入室内作空间上之融和；外观格栅间于模型室位置，嵌挑一玻璃光盒，成为夜间时的视觉焦点，作为销售渲染之媒介及建筑之辨识度。

外接待驻车区，是于外观量体中，内挖之空间，形成外观的层次，并为来访贵宾驻车时遮风挡雨之屏障，刻意全密式之迎宾大门，期望营造当门片开启映入眼帘惊艳气势；入口的接待厅，全景落地观景窗，窗外景观一览无疑，天花及墙材以体现自然质感之风化木为主要素材，气氛的灯光，一组充满居家氛围的家具，周围满布的花艺布置，就是不做太多 商业销售目的的铺陈，只期来宾能感受本案拥揽自然环境的基地特色及优势。

来宾洽谈区以风化木面材、具有量体厚实感之大墙作为区隔，类柱列式的长廊端景焦点映入的即是极富艺术创作意涵的骏马立灯雕塑，呼应本案自然之主题，也将本案未来公共设施中开发商精心为住户注入艺术收藏之用心，作一实现。

在接待空间整体秋香色系的氛围中，纯白色大理石面材之楼梯，活跃地呈现在空间中，似乎在热烈地欢迎来宾造访位于二楼未来的居所样品间；楼梯行进间，自然树艺之装置，又跳脱在视线上，好似窗外之型树，穿越了墙体的限制，进入了室内空间与人为伴。

所有设计的语汇，窗景之形塑，材料的搭衬，色系的选择，其实都是经过长期而细腻的设计思考，但却又不着痕迹地让空间产生富蕴自然生命主题之渲染力，藉而传达本建案基地的优势菁华，丝毫不带浓厚的商业气息，在一自然环抱的场域中，圆满地达成接待中心的实质上销售的任务。

项目名称：九昱十美 接待中心
坐落地点：台北万芳
投资兴建：九昱建设
销售企划：海悦广告股份有限公司
设 计 师：谭精忠
设计单位：动象国际室内装修有限公司
参与设计：赵百怡、何芸妮
摄 影：庄孟翰
面 积：516 m^2
主要材料：镀钛铁件、风化木染色、理石楼梯、特殊漆、满铺地毯

This case is located in the natural mountain scenery of YunShan district, and the base also located in the environment of natural slope. So the natural charm and the home field spirit of the design of the reception center, naturally, focused on the natural environment.

Into the road of the base, according to the original topography, the lateral expansion of the building, showing a broad face, in order to bring out the wide base. Different perspectives reflect different views. We are trying to mix the natural scenery together.

Parking area outside the reception is in the amount of body appearance, the digging of the space, forming the appearance of the level, and the car shelter could be the barrier to get out of the rain for the visiting guests. All deliberately closed of the type of welcome door, It hopes to create a different feeling which could catch the guest's eyes. The entrance reception hall, the panoramic viewing floor windows, the landscape is cleares to be saw outside the window. The ceiling and wall materials to reflect the natural texture of weathered wood, which as the main material. The atmosphere of the lighting, the atmosphere of a full feeling of home furniture, which covered with a floral arrangement around.

The surfaces of the area of guests discussing is used byweathered wood. It has a large amount of solid sense of the body wall,which will be as a segment. It also has the same subject which is to feel the nature.

Surge in the reception room for the whole atmosphere of color.The pure white marble surfaces of the stairs, It actively presented in the space.It seems to warmly welcome guests to visit the future home on the second floor between samples.Between the stairs. It is out of the view and the restriction of the wall.It comes into the space to be the partner.

All the design vocabulary,window view of the shape,color choices,They are actually designed after thinking long and delicate.It want to convey the subject, which is natural life,to achieve the task of actually selling.

SHENZHEN HUAYANG

深圳花样年花郡销售中心

NIANHUA JUN SALES CENTER

本项目由花样年投资公司兴建。“花样年”的公司名称就已经代表了旗下品牌的3个系列。“花”系，主打精品高档城市公寓，比如说花郡项目；“样”系，主打大型居住综合体，比如说成都和深圳的花样城项目；“年”系，主打商业办公综合体。

花样年花郡位于宝安中心区核心区位，是宝安的文化、商业、商务、总部经济和体育中心。可以预期未来花样繁华，花样灿烂的生活将在这里拉开序幕，设计师将花的美丽意向，以现代的设计手法融入到空间中去，带给大家美好的想象。

售楼部外墙包裹着花朵图案的蓝色有机玻璃片，在灯光的照射下洋溢着浪漫气息，营造出繁花如锦的意向。步入大门，净白的接待厅让眼前变得明亮舒缓。纯白色的背景墙，白色大理石的接待台，以及接待台上4盏内壁粉红色的大吊灯，构成了童话般纯美的视界。

以接待台为中心枢纽，左侧为模型展示区及洽谈区，右侧为影音区及贵宾室，人流

项目名称：深圳花样年花郡销售中心
项目地址：深圳大梅沙
项目面积：600 m²
竣工时间：2009年
主要材料：非洲柚木地板、非洲柚木实木、砂岩
设计单位：于强室内设计师事务所

在有限的空间里得到了很好的分流。

从接待台右侧空间步入，经过一个家具陈列墙，各种造型可爱，体量轻盈的家具演绎着生活的情镜。通道空间挑高达两层，因而显得极为阔朗。贵宾洽谈区在这里分为两层，下层半开放式，以圆弧造型的家具和松软的沙发共同构成轻松的氛围。上层为影音区及会议区，以茶色玻璃隔开外界的干扰。

接待台左侧的模型区与洽谈区皆采用开放式布局，便于人群聚合。先是经过白色底座的模型台，客人可以全面了解社区概括。再往前经过一个圆型吧台，吧台中间一条花柄状的造型柱，演绎出花朵的意向，再往里则是洽谈区。洽谈区用了温暖的木色铺陈，营造出其乐融融的洽谈氛围，墙身玻璃上金色的花朵图案呼应外墙的图案，营造出一个如花朵一般绚美的世界。

在这个充满梦幻和童话色彩的空间里，关于家的梦想如此触手可及。

This project is the construction of Huayangnian investment company. "Huayangnian" the company name has represented the brand's three series. "Hua", Department of flagship boutique luxury city apartments, for example, Huajun project; "Yang" series, the main large residential complexes, such as Huayang City project; "Nian" series, the main commercial office complex.

Project is located in Baoan district, is the cultural, commercial, business, economic and sports center headquarters. Prosperous future can be expected, wonderful life will be here.

Sales department wrapped wall flower pattern of blue plexiglass sheet, in the light irradiation was filled with romantic atmosphere, creating the intention of flowers such as Jin. Into the gate, so white and the eyes become bright reception hall. Pure white backdrop, white marble reception desk, and the reception stage four walls pink chandeliers, the pure form of the fairy-tale vision.

Reception center, the left display area and to discuss areas on the right area

and VIP room for the video, people in the limited space has been very good diversion.

From the reception room into the right side, through a display of furniture, walls, furniture and adorable display of life in the air. Pick up two floors of space, which is extremely wide and Lang. Divided into two VIP negotiations, the lower semi-open to the arc shape of furniture and soft sofa together constitute the relaxed atmosphere. Conference for the audio-visual area and the upper zone, separated by tinted glass outside interference.

Model area and negotiations are an open layout of the area, easy walking crowd. After the white model units, guests can fully understand the general community. Then forward a round bar, bar-like shape of the middle column of a stalk, and then further discuss areas where it is. Discussion area with a warm wood color, creating a enjoyable atmosphere to negotiate, on the glass walls of the golden flower design wall echo pattern, creating a brightly world of beauty. Full of fantasy and fairy tales in this color space, so on the dream home within reach.

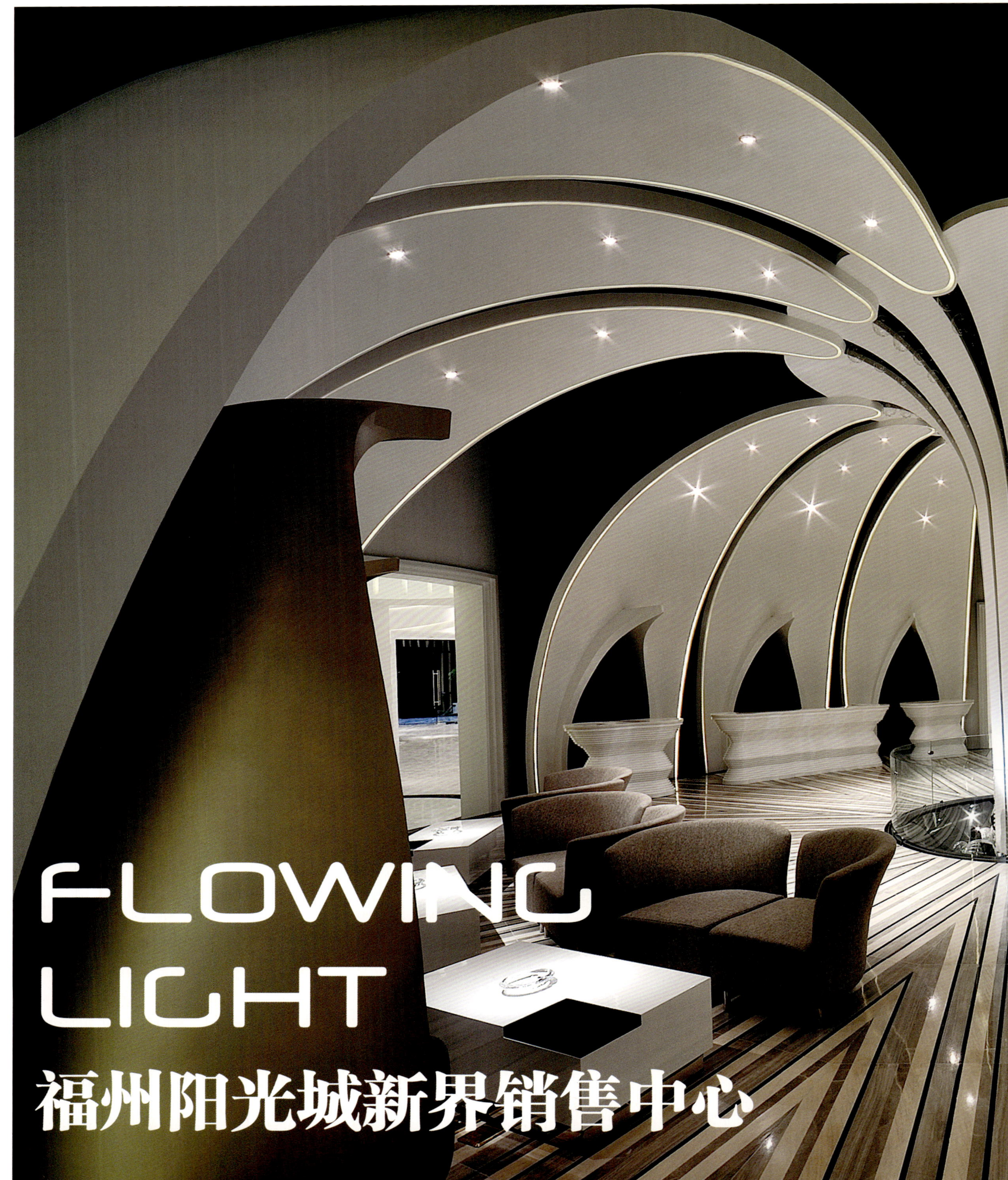

FLOWING LIGHT

福州阳光城新界销售中心

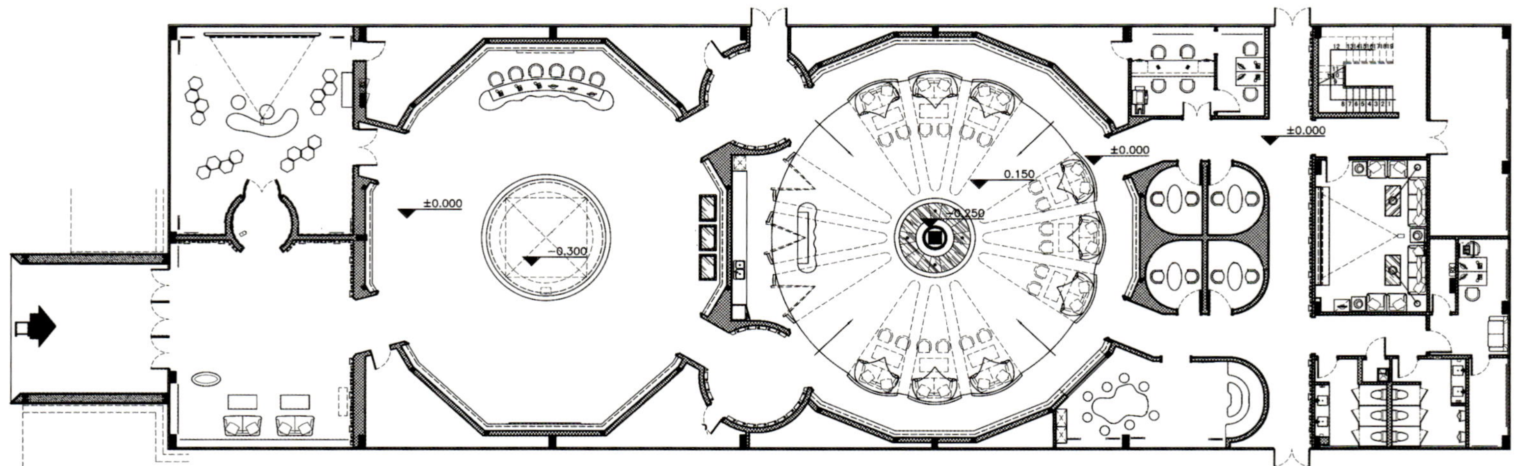

1:100

阳光城新界销售中心坐落于福州金山区，是新兴的居住开发区。项目客户定位为追求时尚的年轻中产、白领。整体的设计思路是围绕光线的感知，并通过不同的形式表达。以此体现人们对光线的不同感觉，或强烈，或柔和。

从建筑方案设计开始强化光线对室内的影响，建筑方案采用彩色覆膜玻璃拼块做法，强化建筑本身远距视觉冲击力，从周围的环境中拖影而出，锁定目标客户群的第一归属感。随光线不同变化，彩色玻璃外墙也会产生不同的视觉效果。色彩的变化也将渗透进室内空间，给内部空间色彩也带来丰富的变化。

入口门厅

穿过厚重的淡金色门廊，进入门厅区感受到竖向光槽产生的强烈的秩序感，吊顶中心环形垂吊了大量铜管形成空间核心，竖直的垂吊方向和墙面的灯槽方向一致，形成模拟光线垂直向下发散的形态。阳光透过外幕墙彩色贴膜玻璃，形成的彩色光晕映射在白色墙面和地面上，空间充满丰富的色彩变化。

在通往影音厅圆形过厅地面上设有互动投影，使用地面的互动投影是为了把新进入的参观流线引入影音厅，减少门厅处人流进出的交叉流线。

影音厅

影音厅内没有了强烈的竖向线条的冲击，吊顶散布的蓝色LED光源慢慢地变换亮度，四周墙面设计有暗色软包。没有视觉的焦点，使参观者的心绪平静下来，清晰的巨大的投影幕的内容成了新的视觉焦点。

沙盘区

穿过影音厅暗门，进入沙盘模型区，在近似圆形的围合空间内墙面环绕曲线光槽组，虚化空间边界感，使得空间充满虚空感，不受边界所限制。强化光线在空间外围墙面的流动效果，从曲线光槽组中精心选取水波形透光面将室外阳光引入室内，光线给白色曲线墙面带来了丰富的色彩变化。曲线墙面有限的水波形透光带，给墙面增加了些许彩色光线的流动感，使得墙面既不刻板，又不会喧宾夺主。沙盘采用半下沉式设计，使得参观者视线与沙盘模型形成最佳角度。沙盘侧立面也做了精心设计，采用锥形折面灰镜饰面，计算好灰镜折射角度与法线角度，以参观者视线高度看向锥形侧面有向沙盘下的无限延伸感。中心沙盘与吊顶中心呼应，采用了灰镜材料，与外围白色曲线墙面形成强烈灰度对比，吸引参观者注意力，强化中心沙盘的重要性。白色三维弧形线条接待台与曲线墙面很好地融合在一起，达到了虚化周边强化中心的目的。

洽谈区

洽谈区四周多边形曲线光槽组继续营造无边界的虚幻感，中心大胆设计出巨大"花瓣"造型将流动的光线静止住，"花瓣"造型边缘设计了极细的光槽，光线由上而下包围每个洽谈组，花瓣结合中心的倒锥形"花心"造型形成了完整的抛物线，弱化了空间转折，地面的中心放射状线条形式，强化了造型的视觉冲击力，在参观者视角形成震撼的透视扭曲效果。沙发采用壳体结构形成半围合形式，声波被壳体结构强化并限定折射范围内，最大程度降低人们交谈时相互的声音干扰，壳体结构外墙面使用淡金色金属漆，在光线漫反射下给人沐浴阳光的感受。

设计师为销售中心引入光线的概念元素，运用一系列手法，由外到内，由流动到静止，由虚幻到切实，由塑定空间到伸张扭曲，利用光线在各种造型及材料质感上呈现的变化，竭力强调了空间个性。

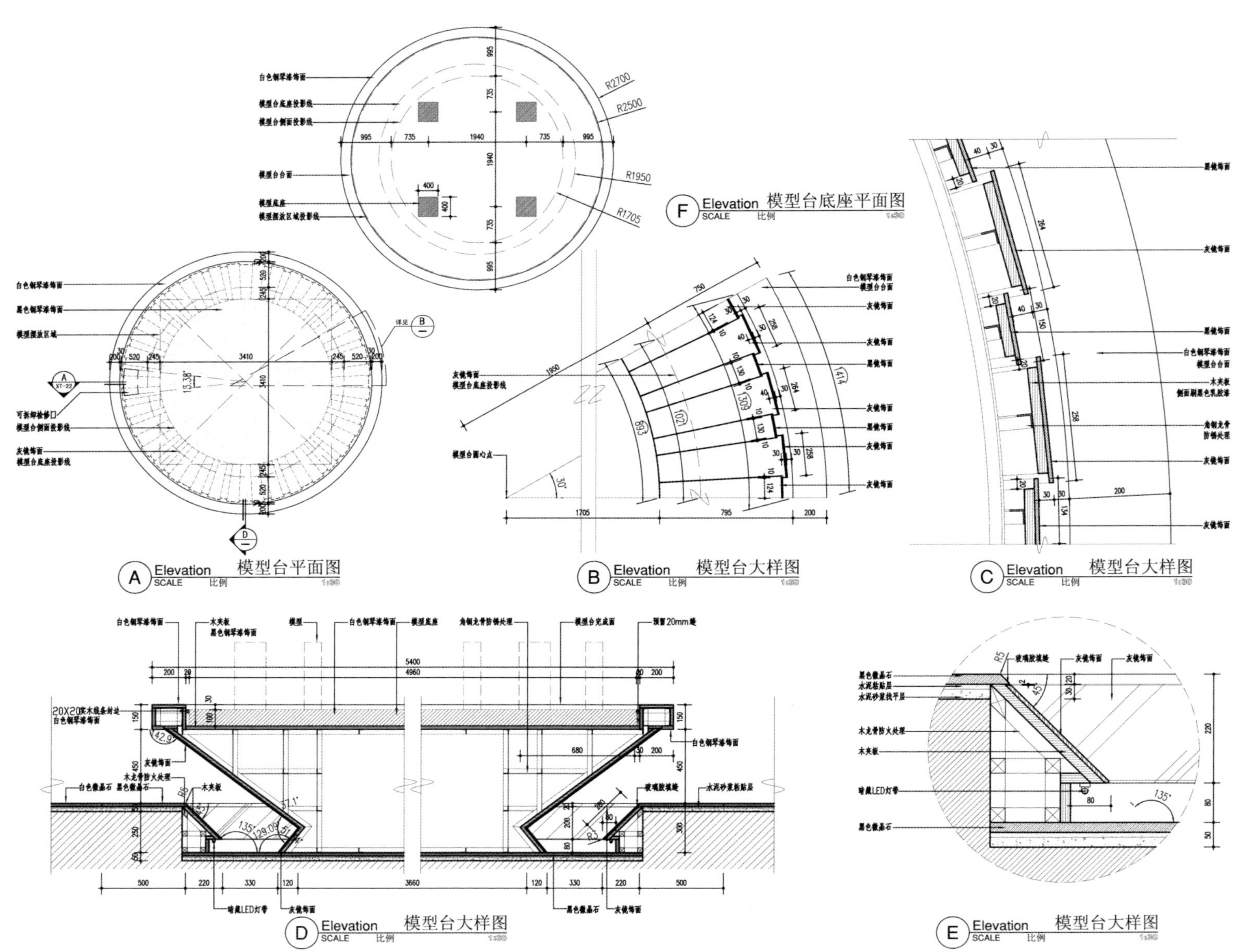

F Elevation 模型台底座平面图 SCALE 比例 1:30

A Elevation 模型台平面图 SCALE 比例 1:30

B Elevation 模型台大样图 SCALE 比例 1:30

C Elevation 模型台大样图 SCALE 比例 1:30

D Elevation 模型台大样图 SCALE 比例 1:30

E Elevation 模型台大样图 SCALE 比例 1:30

Sun City sales center is located in Fuzhou Jinshan district, it is a new residential zone. This project is fit for young middle-class, white-collar workers who seek to the fashion. The overall design idea is expressed around the beginning of light perception. Refining theme: the feeling of light is calm, flowing, soft, bright and colorful.

From outside to inside: Designers strengthen the influence of light when they began to design. The building solution adopted the Color coated glass patchwork to enhance the visual impact, makes the construction stand out from the surrounding environment immediately, giving the target customer the first sense of belonging.

Entrance: Through dense pale golden porch, in the lobby area, visitors feel a strong sense of order produced by vertical light slots, the center of ceiling hangs many copper tubes which form the core of space. The direction of the tubes is as same as the light slots, making visitors feel that the emission of light is vertical.

Media Room: First experience in the video room is not a strong impact of the vertical lines, there is no sense of order produced by the direction. Light of the blue LED in the ceiling slowly change the brightness. The surrounding walls are decorated with dark soft materials to weaken the sense of siege. There is no visual focus, the mood of visitor is comfortable and clear content on the huge screen has become a new visual focus.

Sand table area: Through secret door of the media room, visitors get into the sand table model area. By using curve light slots surrounding the nearly circular space, boundaries of space are blurred, so in this area visitors can't perceive boundaries because of the space filled with empty feeling. Strengthen the Liquidity effect of the light. Introduce the sunshine from outside to inside by carefully selecting transparent surface from curve light slots. Sunshine brings a wealth of color changes to the white curve wall.

Discussion areas: Around discussion areas, there are some polygonal curve light slots which continue to create a feeling of no boundaries. The Center boldly designs a huge "petal" shape which makes the sunshine look like static. Around the "petal" design some tiny slots. Each of discussion group is surrounded by the top-down light. The inverted cone "flower heart" with "petal" forms a complete parabola, making space look gentler, and the form of radial lines on the center of the ground strengthens the visual impact of the shape. The distortion effect gives visitors a stunning feeling.

Designers introduce the concept of light for the sales center, using of a range of techniques, from outside to inside, from mobile to stationary, from unreal to real, from formed spaces to the twisted ones, by using changes when light meets different shapes or materials, designers strive to emphasize the perceived personality of spaces.

阳光城·新界
COLORFUL LIFE

中骏天峰会所
ZHONGJUN TIANFENG CLUB

设计围绕着如何在长 21m 高 7.6m 的局限性空间中满足销售会所的刚性功能需求的前提下，通过造型灯光、色彩、材质等设计需要来彰显主题定位（摄政王钻石）的空间气质。

Under the guideline of realizing the basic functions of the sales salon in a limited space which is 21m high and 7.6m long, this design demonstrates the noble quality of the theme (Regent Diamond) through special combination of the lights, colors and materials.

设 计 师：黄振耀

设计单位：东峻设计顾问有限公司

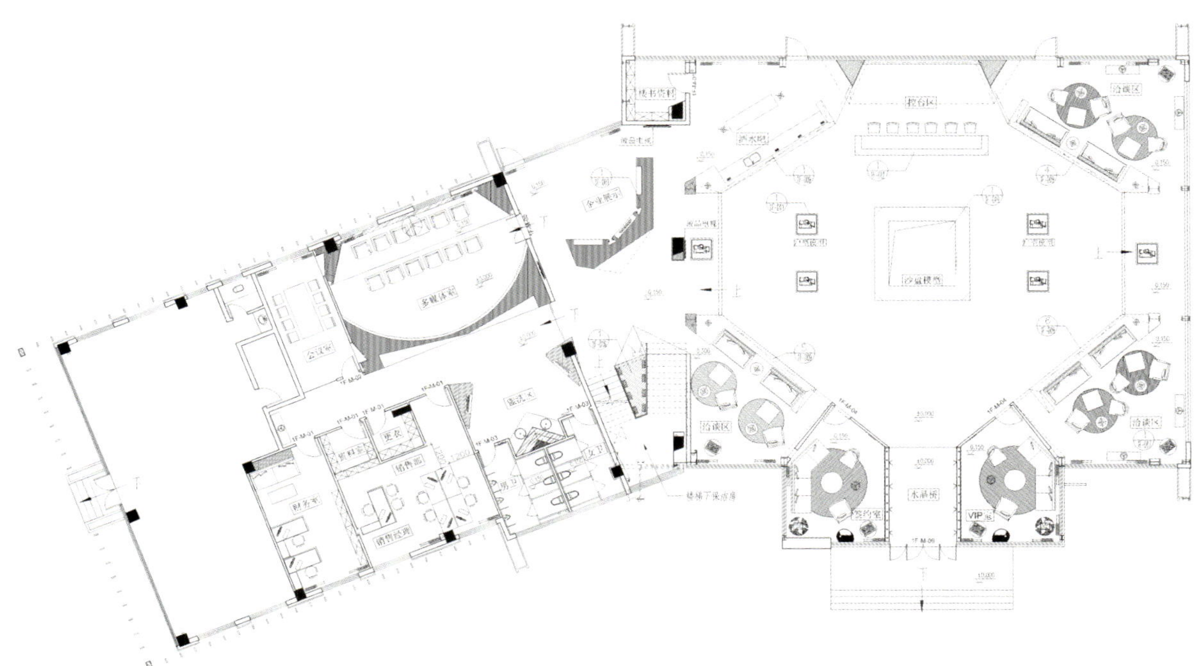

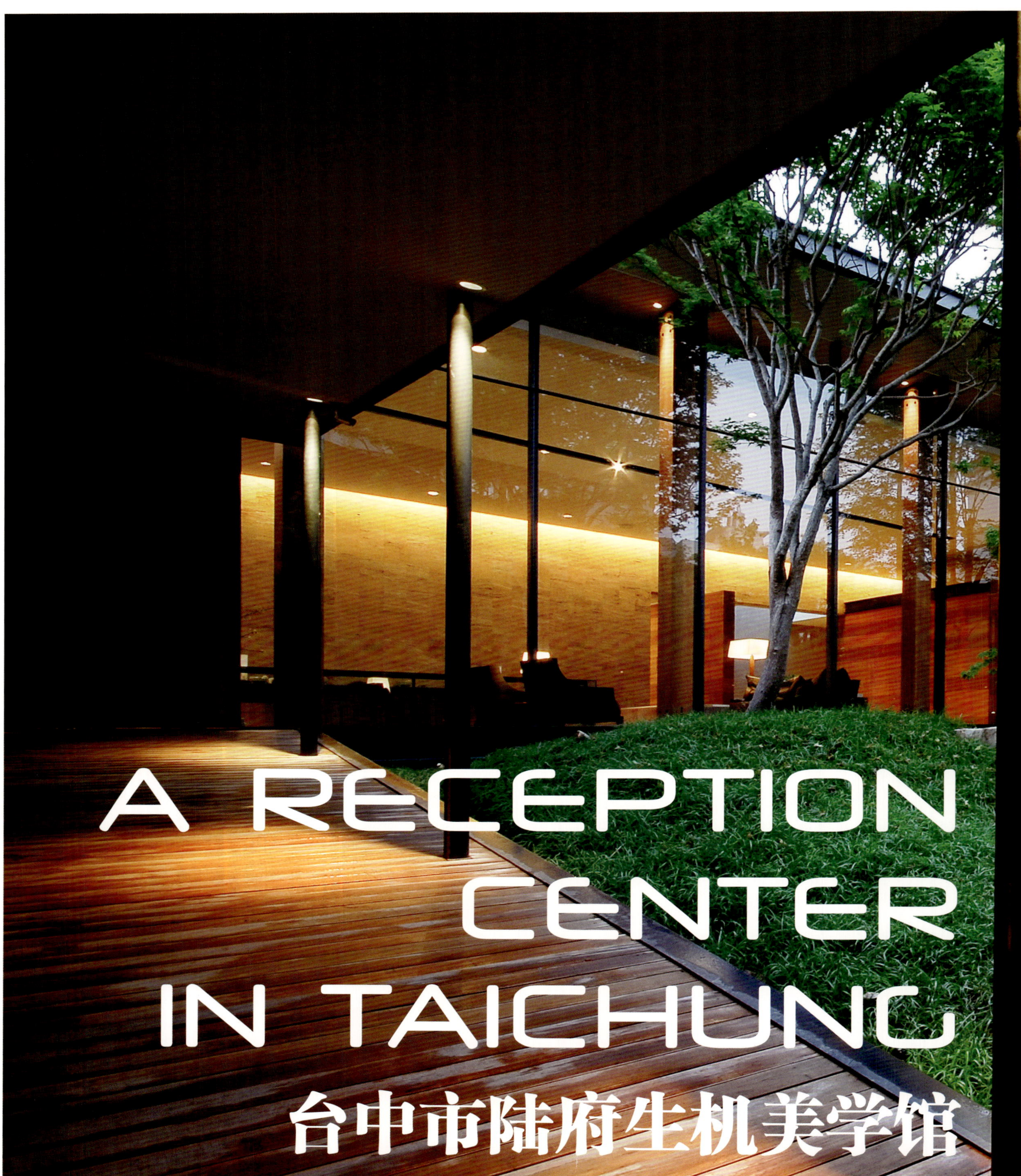

A RECEPTION CENTER IN TAICHUNG

台中市陆府生机美学馆

陶渊明笔下的桃花源距离我们很远，也可以很近；它远时在天边，近时却藏在每一个人的内心里。有人说建筑要符合一个地方的特色，人们要能在建筑中看到自己。那么在以桃花源记为主轴规划设计的本案中，或许我们可以寻访到一些设计师心中那座桃花源的影像。

从本案的平面图来看，无论狭长的入口部分，还是看似随意的斜坡设计都隐约地透露出一丝中国古典园林的韵味。把园林设计的意境提炼升华，将其运用在现代建筑中，是这个设计的一大特色。设计师陈天助先生对中国园林研习多年，用他自己的话来讲，在设计中也不免会无意识的将园林设计的一些手法运用进来。

在设计中，设计师希望能够围绕主轴营造出园中游园的绿意，在与邻近的丰乐公园彼此相互呼应的同时，也印证着业主提出的"绿海工程"这一主题。想要在城市中提供给人们一个心静自然、天人合一的环境也是设计师最初的想法。这间售楼处布局精简质朴，却不失精致典雅的美感。由于空间较小，设计师希望能够运用在水平和垂直两个方向所做的穿插结构来达到一种移步换景的效果。他没有企图把不同的功能空间细胞冻结在一个封闭立方体中，相反的，通过自己的设计手法将建筑的高度、宽度、深度与时间这个设想性的思维整体，在开放空间中使其以一种接近于全新的塑性表现出来，这样的设计具有一种或多或少的飘逸感，出离了一般建筑本身的那种厚重。而那仿佛自然形成的坡道就像是园林中曲折迂回的廊道，在不断的变换中与内部的环境直接对话。

虽然建筑的性质是接待中心，但当人们步入这个空间的时候总会有游走于艺廊之中的感觉。建筑空间呈现出一种连续的、逐渐的、复杂的、精致的变化，足以让人忘记这里的商业氛围，而这也是与业主想要打造社区人文工程的初衷相吻合的。

整个接待中心的设计跳脱了传统的以展示坪效为重心的考量，取而代之的是让人们感觉温暖、安静的空间体验，处处精彩，并且让宾客们对此处的绿色环境产生共鸣。步入藏而不露的入口，透过园林式的回游路径，设计师以矮低天花板、窄墙塑造空间，当人们的步伐穿透中庭，便可以逐渐感受豁然开朗的空间层次，参观的动线引导人们行至主会馆、样品屋、洽谈区，整个设计显得自然而不造作，运用宁静的氛围构筑空间张力，让隐密的入口与通畅的内庭构成明暗与宽窄对比，产生韵律，收放自如。

建筑的一些墙壁上开凿了大小不一的

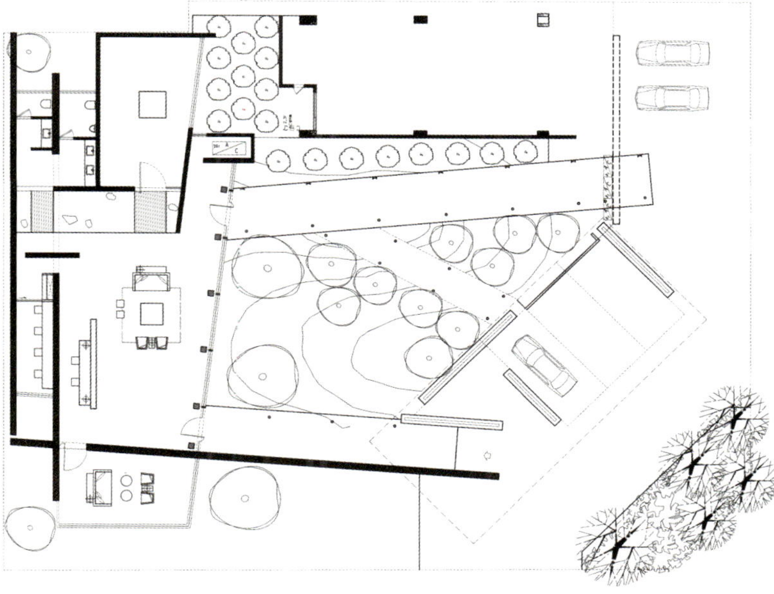

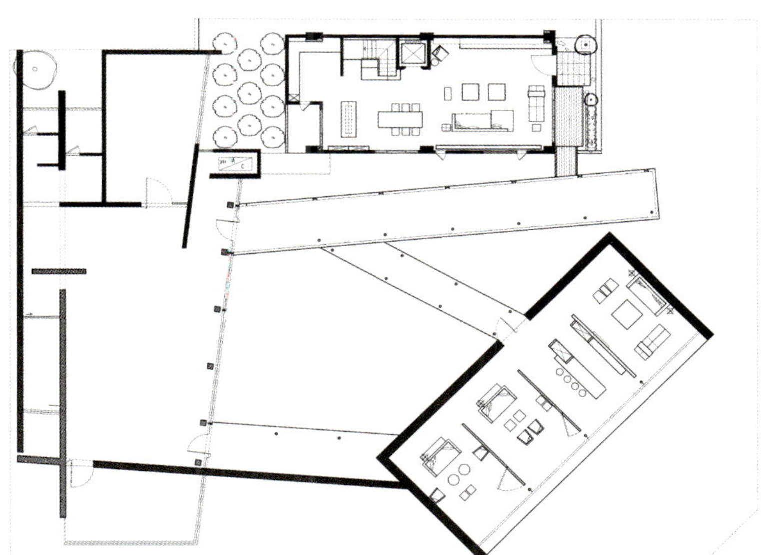

墙洞，则是设计师对园林中花窗形式的一种提炼。中庭的坡道部分设计师本来是要做成“岭”的效果，却意外的做出了“峰”的感觉，第二个坡道较长也引人走上“峰”顶，这同时也意喻着业主蒸蒸日上的事业。

人们站在二楼，就可以直接望到对面的公园，在视线上没有任何的阻碍，从而形成了某种形式上的借景的效果。也可以说整个基地的自然条件决定了这样园林上的布局，在围与透，虚与实的交错中给人们一种放松、愉悦的感觉。

建筑选用了质朴的石材、实木等天然的材质，这些都是最名贵的大理石与进口木材，在造型方面并没有任何花俏之处，却隐隐透露的是一种低调的华丽质感。值得注意的是建筑的支柱都是选用缅甸柚木而非钢骨。天然材料的运用一方面是考虑到接待中心作为一种临时的建筑，到了一定的时期便会被拆除，天然的材质可以重新被回收利用。而另一方面天然材质的朴质自然与精致的定制家具构成粗犷与细腻的对比，也让空间有了另外一种不同的味道。

The paradise described by Tao is very far from us, It also can be close.It is far in the horizon.It is near to us which is just like be hidden inside of every people's heart. Some people say that building should have their local characteristics.People could be able to see themselves in the building. Well, the paradise described by Tao as the main planning and design of the case.Perhaps we can look for the image of paradise from the hearts of some designers.

From the plane of the view of the case.Whatever the part of the narrow entrance or seemingly random slope design are subtle, revealing a trace of the charm of Chinese classical gardens. Artistic Landscape Design refining the sublimation of its use in modern buildings, is a major feature of this design. Designer Mr. Chen Tianzhu study of Chinese gardens for many years, the words of his own, All can not help in the design.The design of the garden will be unconscious of some techniques in the coming.

In the design, the designers hope to create a garden around the main shaft of the green garden, park in the adjacent Fengle echo each other, but also confirms the owners of the "Green Sea Project" of the subject. They want to provide to a calm, natural, nature and man's environment to the people which is the designer's original idea.The layout of this streamlining of the sales office is simple yet refined and elegant beauty. Because of the space smaller, designers hope to use in the horizontal and vertical directions interspersed structure made to achieve the effect of a venue for King. He did not attempt to freeze the different functional cells in a closed space cube.On the contrary, through their own way of design of the building height, width, depth and time that this idea of thinking as a whole, in the open space to make it a shown close to the new plastic, this elegant design has a more or less sense of a general building itself out of the kind of heavy. And that seemed like a natural formation of the ramp in the garden paths winding corridors. They have directly internal dialogue in the changing environment with constantly.

Although the construction of the na-

ture is reception center, but when people enter the space. When there is always the feeling of being in the gallery. Architectural space takes on a continuous, gradual, complex, delicate changes, enough to forget the business climate here, which is the human community and the owners want to build the original intention of the project to coincide.

The whole design of the reception centers escaped the traditional way which shows floor as the focus for consideration. Instead, it is to make people feel warm, quiet space experience, always wonderful, and let the guests on the green environment resonate here. Into the hidden entrance to the migratory path through the garden, the designer in order to short the low ceilings, narrow wall of mold space, when the pace of the people through the atrium, they can gradually become clear space experience level, visit Fixed line and guide people to the main hall line, sample house, discussion areas, the whole design is natural and not artificial, the use of space to build an atmosphere of quiet tension, the entrance to a hidden form in chambers with smooth shading and width compared to produce rhythm, to close place easily.

Some of the walls of buildings of different sizes drilled a hole in the wall, it is the form of a flower extract of the garden window by the designer. Some of the designers in the atrium of the ramp was supposed to make "Ridge" effect, but the accident made a "peak" feeling, but also the introduction of a second long ramp to a "peak" top, this also means the flourishing career of the people.

People standing on the second floor, glancing across to the park directly in the line of sight without any obstruction, thus forming a sort of by the King of the results. It also says that the entire base of the natural conditions on the layout of such a garden. It gives us a feeling of relaxed and enjoyable.

Choosing a rustic stone building, wood and other natural materials, these are the most expensive marble, and imported wood, the shape does not have any fancy terms of the department, to some extent, revealed a low-key gorgeous texture. Note that building the pillars of Burmese teak is used instead of steel. The use of natural materials while also taking into account the reception center as a temporary building, at a certain period of time would be removed, the natural material can be recycled again. On the other hand, the primitive nature of natural materials and exquisite custom furniture and exquisite form rough comparison, but also to space has been on a different flavor.

RECEPTION CENTER OF MIDTOWM IN TAIPEI

台北翔誉之心接待中心

工程名称：台北翔誉之心接待中心
坐落地点：台北市南港区
面　　积：1F~285㎡，2F~440㎡
设计公司：齐物设计
参与设计：林咏淳、高泉瑜、张亚琁
主要材料：黑色洗石子、清玻璃、黑铁、橡木染色、雪白银狐、皮革、茶色玻璃
设计时间：2009.12 ~ 2010.02
竣工时间：2010.06
摄　　影：卢震宇

在房地产业蓬勃发展的今天，售楼处的设计往往承载着传递整个楼盘设计理念、居住氛围，人文环境的重任。位于台北南港区的翔誉之心楼盘，策划者就希望通过售楼中心的室内外设计来诠释一个“森林”的概念，以此来提升楼盘自身的价值，推广理想人居的概念。

但要用钢筋水泥来塑造具有生命感和自然感的森林，对于设计者来说是个极富挑战的工作。然而在本案的设计中，设计者通过独具匠心的建筑外立面设计、材质的巧妙运用，以及多元化的表现手段，将森林的概念成功注入这样一个硬质空间，获得了良好的视觉和感官效果。就外观而言，整个售楼中心的主体被笼罩在仿照树形设计的镂空外壳中，创造了一个过滤光阴的界面，树的元素被放大和戏剧化，让灯光璀璨的内部空

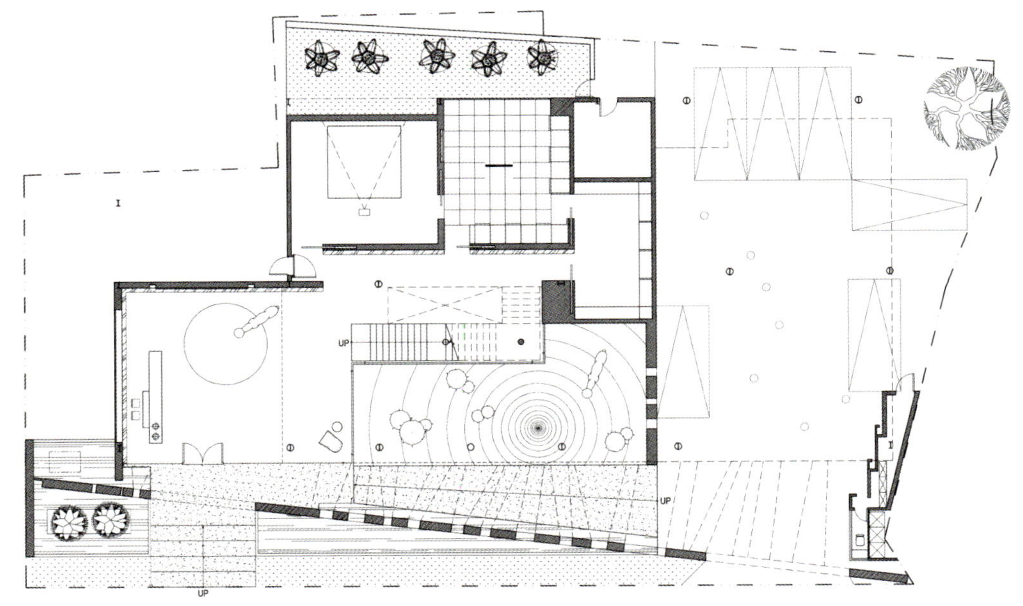

间仿佛掩映在婆娑的树影之间，若隐若现，引人入胜。入口处的一棵真树成了整个建筑的LOGO，使进入此地的人们从建筑的外观就感受到扑面而来的清新气息。走入大厅内部，整个建筑曲折迂回的墙体和划分空间的隔断都选用了木质材料装饰表面，并利用木材天然的色泽和纹理以横向空间为装饰主轴，塑造了富有层次感的墙面肌理。木质天然质朴的触感，配以温暖的黄色灯光，四壁仿佛散发着木材的馨香，整个空间充满家的温馨和宁静。

在室内陈设和家具的选择上，设计师不断汲取森林中的各类元素，或具象的还原，或抽象的模拟，万变不离其宗，始终围绕着森林这个主题进行创作。例如一楼偏厅的地铺设计中，设计者将木材和玻璃做成相间的方格，并在玻璃面下设计了仿真草坪，利用视觉的传达，使人们好像脚踩在柔软的草地上一般。还有随处可见的草坪地毯，moooi house lap，仿真鲜花。特别是漫飞在整个空间中的亚克力蝴蝶装置，它是设计师以点对点的模式严谨的从设计图纸上定位还原在建筑空间中的，它们按照理性的数理关系阵列式的悬挂在空中，从一楼延伸至二楼，将空间用同一种元素串联在一起，赋予整个硬质空间灵动的生命感，森林的诗意情致跃然于眼前。建筑内庭的空间里，设计师用未经任何加工的原木以同心圆的形式排放在地上作为地铺，生动模仿了树木的年轮。伫立其上的小马和怒放着鲜花的仿真树，在真实与虚幻间讲述着关于森林的童话故事，充满蓬勃的生机。包括在空间整体色彩的把握上，也保持了自然简朴的中性色调，原木色、黑色、白色、深灰色、草绿色，这些都是自然界的色彩，没有过多的对比和渲染，只是用光影的变换来增强色彩的梯度，呈现森林郁闭空间里多变的光与色。

这是一个关于绿色和生命的设计，由森林这个词衍生出的一系列元素被完美融合在一起，在这个钢筋水泥的世界里展现了一幅自然美景。

In this booming days, the sales office design often carries transfer the real estate design, living environment, human environment of the task. Nankang District in Taipei Xiang Yu the heart of projects, planners hope that through the sales center for interior design to interpretation of a "forest" concept, in order to enhance the value of its real estate to promote the concept of good living.

But the use of reinforced concrete to create a sense of life and the natural sense of the forest,The designer is a challenging task. However, in the case design, the designer through the distinctive facade design, clever use of materials, as well as a wide range of performance measures, the concept of success of the forest into a rigid space, access to a good visual and sensory effects . For the appearance is concerned, the main body of the sales center is shrouded in the design of the hollow shell modeled on the tree, they created a filter time of the interface, the tree is enlarged and dramatic elements, so that the interior space of Light and whirling like the shade of between the shadows of the trees, indeed, fascinating. Become a real tree at the entrance to the building of the LOGO, so enter here the appearance of people from the building to feel fresh air blowing. Into the hall inside the building by tortuous roundabout space partition walls and are equipped with a decorative surface of wood materials and use of color and texture of natural

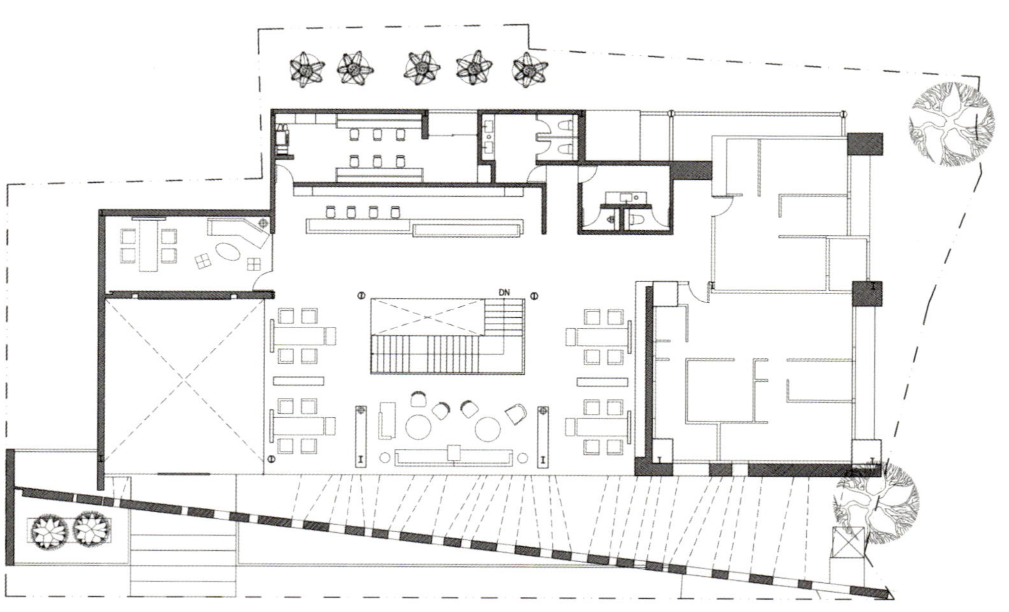

wood with decorative spindle horizontal space, create the rich layering of the wall texture. Simple touch of natural wood, together with the warm yellow light, walls of wood exudes a fragrance as if the whole room full of family warmth and tranquility.

In the choice of furnishings and furniture, the designers continue to learn the various elements of the forest, or the reduction of concrete or abstract modeling, aim, always revolved around the theme of the creation of forest. For example, the ground floor first floor Painting design, the designer made of white wood and glass box, and the glass surface of the lawn to design a simulation, the use of visual communication, so that people like foot on the soft grass in general. There are the ubiquitous lawn carpet, moooi house lap, Simulation of flowers. Particularly diffuse in the whole space flight butterfly acrylic device, which is the pattern designer strict point to point from the design drawings to locate space in the building restored, they Mathematical relationships according to reason is hung in the array air, from the first floor extending to the second floor space with the same element in series, giving the hard life of a sense of space Smart, forests fantastic flash on the front of the poetry. Court building space, the designers used without any processing of logs in the form of concentric circles on the ground as the ground floor emissions, vivid imitation of

the tree rings. Pony standing on it and flowers in full bloom in the simulation tree, between reality and illusion in telling fairy tales about the forest, full of vigor. Included in the overall color of the grasp of space, but also to maintain the natural simplicity of neutral colors, wood color, black, white, dark gray, grass green, these are the colors of nature, not too much contrast and rendering, but with light transformation to enhance the color of the gradient, showing a variable forest canopy space, light and color.

This is a design on the green and life, the word derived from the forest.A series of elements are perfect coming to together. It shows a picture of natural beauty in this world which is full of steel and concrete.

THE WEALTH CENTER OF CHINA SCE PROPERTY

中骏财富中心

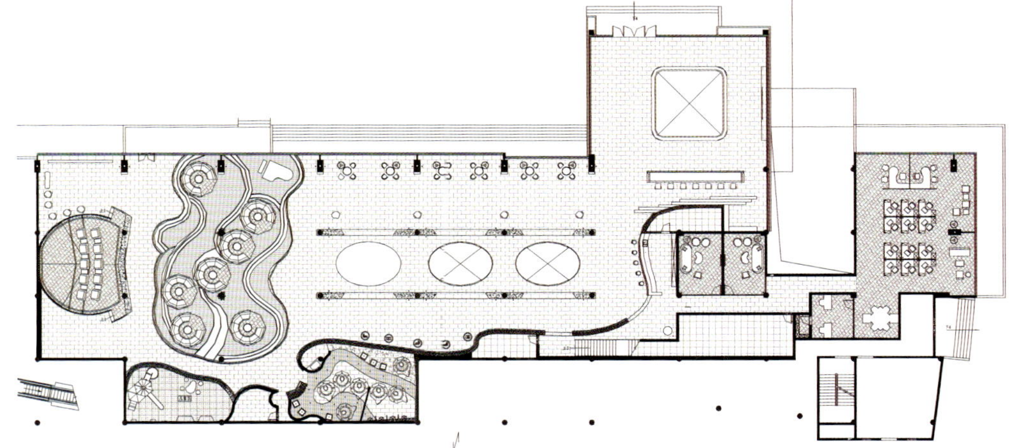

房产销售中心的空间设计以功能性为主要目的。如何从形态和空间上脱离传统销售中心的空间模式，是设计师在本项目中致力突破的环节。本中心的设计颠覆了单一空间概念，在合理分配使用功能的基础上，将“水”这一设计元素延伸性地融入空间的塑造，营造出未来与现实并存的超现实魔幻意境。通过“水”，来表达祥和富贵之意以及人与自然和谐共处的绿色人居概念。

The space design of a Real Estate Sales Office mainly focuses on its functionality. In this project, designers would be dedicated to make breakthroughs on pattern and architectural space which are totally different from spatial pattern of traditional sales offices.

The unique feature of the design of this Wealth Center as a Real Estate Sales Office lies in its subversion to the conventional concept of mono-space. On the basis of reasonable distribution of functionality, the core design element---Water would be fully considered and integrated in the design of architecture space. In this sense, a poetic imagery of hyper-reality would be deliberately constructed.

The core design element—Water symbolizes two meanings; on the one hand, it means auspicious and prosperous; on the other hand, water is an organic link between nature and man, in this way, the green living concept of the harmony between human and nature could be excellently expressed.

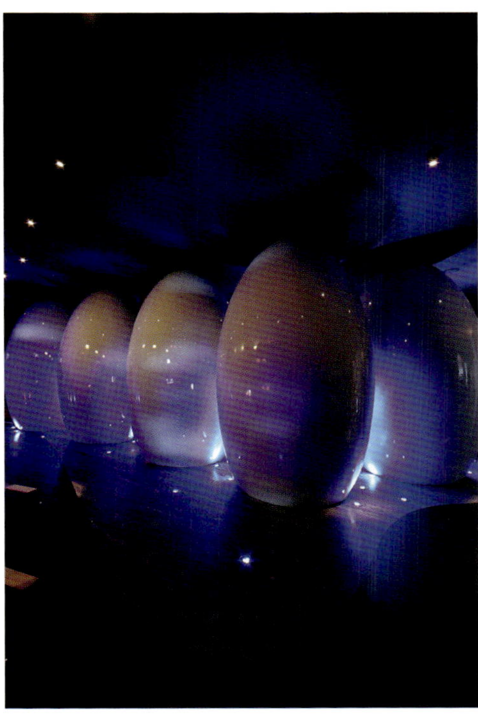

GALAXY SOHO SALES CENTER

银河SOHO销售中心

项目名称：银河SOHO销售中心
坐落地点：北京
设　　计：扎哈•哈迪德
摄　　影：雷坛坛

银河SOHO是占地50 000 m²,总建筑面积328 204 m²,集商业办公于一身的大型综合项目。这座融动的优美建筑群不但营造了流动和有机的内部空间,同时也在与此毗邻的东二环上形成了引人注目的地标性建筑景观。银河SOHO销售中心的设计是建筑设计语言的集中体现,扎哈·哈迪德的标志:流动性、夸张的造型语言等在此得到表达。

SOHO is an area of 50,000 sqm of the Galaxy, with a total construction area of 328, 204 sqm. Commercial, office in a large integrated projects. This will not only create beautiful buildings within the flow and organic space, but also in the formation of the East Second Ring impressive landmark landscape. Galaxy SOHO sales center design is the concentrated expression of architectural design language, Zaha Hadid's logo: mobility, exaggerated shapes are expressed in this language.

GALAXY
银河SOHO

GALAXY

雄都
遠新
金華苑

THE RECEPTION CENTER OF YUANXIONG JINHUA

远雄金华苑

创造出“精品住宅”的建筑符号

位于台北市内湖五期重划区的“远雄金华苑”建案，是远雄建设以“美学住宅”的概念推出的新代表作。不同于以往的是，远雄在内湖将以“欧陆贵族风”分批推出一系列建案，试图尝试接待中心“品牌形象化”，使人们一眼就能识别出“这是远雄的建案”；同时要求接待中心能在此“长期使用”，其内装风格能兼容之后推出的各系列欧陆艺术风格。因此，玄武设计在设计“金华苑”接待中心时，必须遵循企业主的设计规范，以标准色系、标准化组件来呈现“精品住宅”的形象，还要能运用这些规范，打造出与欧陆贵族风吻合、又能闪露出俄罗斯风华的接待中心。

远雄的建案都是走高价位的住宅，接待中心的设计首要目标就是表现出“国际精品”的形象。玄武设计取法于国际精品设计，将远雄企业LOGO当作基本符号，在建筑外立面做大面积堆栈，组合成一座标志性的立塔，简洁典雅的符号有着清晰的辨识效果，如同国际精品的经典款花纹，创造出属于企业的经典图案，此一图案也将成为远雄接待中心的标准化元素。在接待中心入口的立面橱窗，再次铺陈LOGO符号成一面企业形象墙，此一墙面的材质为镜面玻璃，镜面上以喷砂的方法涂绘出远雄LOGO图案，夜间光线可从未涂绘处透出光线，营造出璀璨的光雕效果，好似一个晶莹剔透的空间，立即给人精品与高贵的感受。

俄罗斯的华丽：金色的辉煌

接待中心的空间布局和功用，则依据标准化的功能分区和展示组件来配置，基本上采用“简约留白”的设计原则，好让空间依据实际销售之所需做弹性的使用，加入其他欧陆风格也不显突兀，符合业主长期使用之所需；但如何在简约的设计中还能带有俄罗斯的贵族奢华风，则是本案的最重要课题。若仿照俄罗斯皇宫中大量使用金色的装饰和家具，以现代人的眼光来说，会是视觉上的疲累和空间中的累赘，玄武巧妙地运用金黄色光线和金色的点缀，在洁净的空间中，恰当地加上属于俄罗斯的金色，非常成功地表现出俄罗斯的主题。

在接待大厅通往主模型区的信道上，以俄式建筑的特点“洋葱头屋顶”的造型来做门框，还有宫廷立灯型状的门窗，让后方的水晶灯从后方晕染出金色的轮廓；二楼洽谈区白色的隔屏，就是一组刻上华丽的水晶灯图案的灯具，透露出金色光芒，烘托出华丽的气氛；当然属于皇室的金色家具是不可或

缺的，但不可太过，于是在窗帘上、家具布料中、织锦上以各种金色色系的交叉使用，如协奏曲般歌颂出俄罗斯的金色年代。

俄罗斯余韵：细致醇厚的白与黑

地板的黑白棋盘图样，是直接仿效俄罗斯圣彼得堡夏宫的下花园中的黑白棋盘铺面，黑的高贵和白的典雅，更加显出皇室的金色尊荣，以白黑为基调的空间设计，成了显耀出金色的最佳背景。

接待中心最精彩的部分，就是在接待中心中轴处，形塑出精品形象的纯白色图案墙；圆柱几何形的图样从地面堆栈至天花板，是由特制的纸筒一个个以人力组装而成的浩大工程，没有任何框架包覆和支撑，纸筒间也没有任何卡榫与崁接，全赖力的平衡和重力的平均分散，保持每个正圆均无变形，最环保的纸也能做为素材玩出空间的特效，堪称玄武设计的创意和创举。纯白的空间加上从外引入的自然光，宛如一个崇高圣洁的仪式性空间，将俄罗斯皇室教堂的空间感，以当代的设计手法与以转译，少了些宗教的肃穆，依然古典雅致，却多了些当代的写意与诗意。本中轴也是串联其他空间的主廊道，无需在每个空间中用力造作出欧式贵族风，通过此空间既能提纲契领式地点出「欧陆精品」的质感，能表现出俄式的华丽风情，也能衔接到欧式的优雅与精致。

在西式调酒中有很知名的黑色与白色俄罗斯，以俄国伏特加为基底，加上香醇的咖啡利口酒和滑润的鲜奶油调和成的鸡尾酒，香甜却后劲强烈；玄武设计在远雄金华苑的设计表现，就像是一个高明的调酒师，以黑色的深沉醇厚、白色的精致绵密来调和俄国皇家的金碧辉煌，而且当黑与白纯到极致时，也是另一种亮眼的金光，余韵却更加深长。

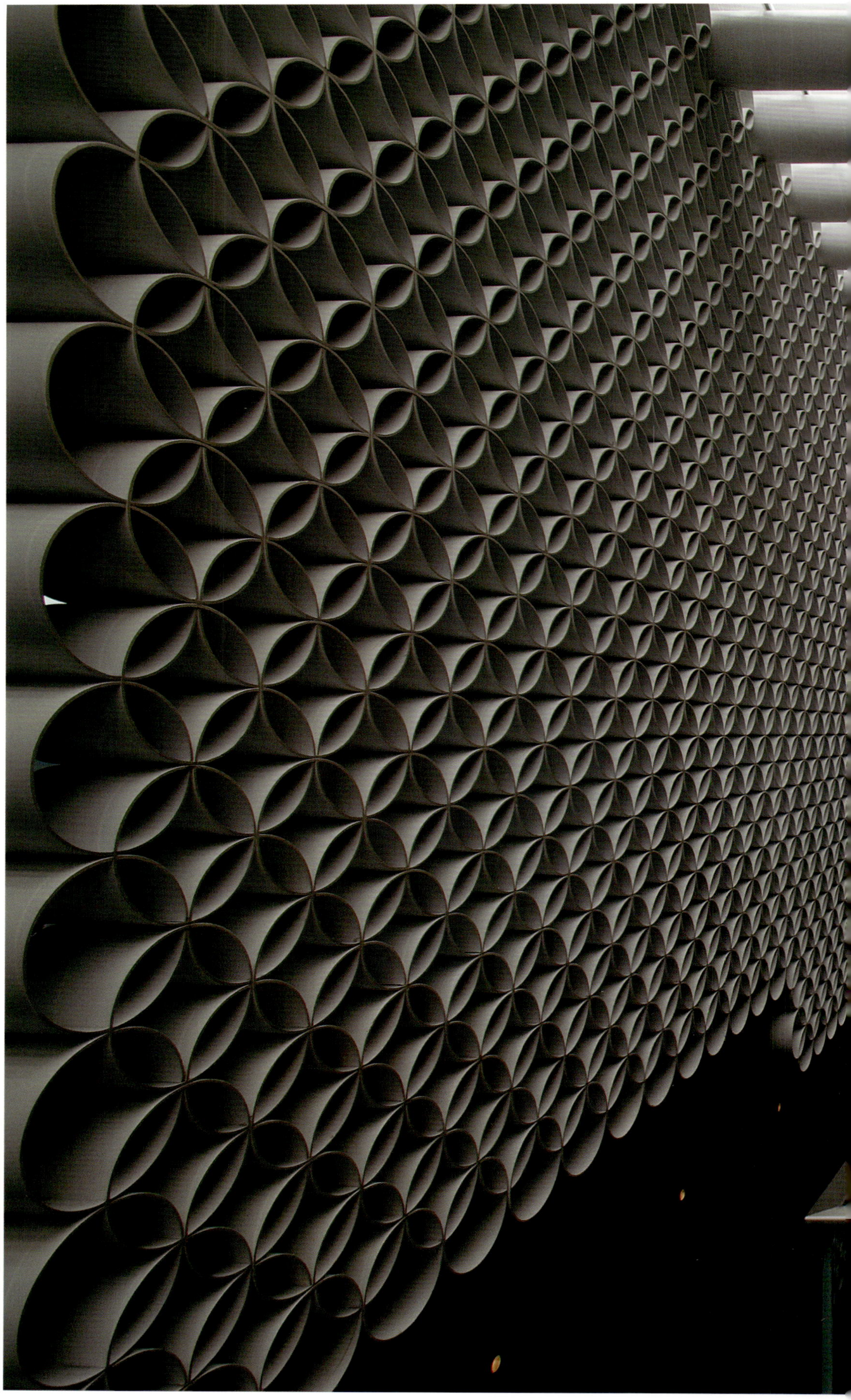

The architectural symbols of Creating a "fine house"

It is located in Neihu, YuanXiongJinhua, Taipei. That is a new masterpiece of the concept of "aesthetics of residential". Unlike to the past.The theme of "European aristocratic style" will be building a series of program.So that people can identify it at a glance.also asked reception center to use it for a long-term.The interior style is compatible with the series after the introduction of continental style. Therefore,we just want to present the reception center of russian elegance.

Buildings are higher-priced.The reception centers, the primary design goal is to show "international quality" image. We want to combine all into an iconic towers standing. Simple and elegant symbol with a clear identification results.As the international boutique Classic patterns, To create the classic pattern of the enterprises.this pattern will also be the standardized elements. The facade windows of the reception center on the entrance.We could present a LOGO symbol wall.This wall is made of mirror glass.It will coated with the LOGO image.In the night, lignt will give us a new feeling which is elegent.

Russia's magnificent: the golden brilliant

The function of reception center.We use"simple blank" to be the design principles.so spaces could be more flexible to join other Europeanl luxury style.If modeled on the extensive use of golden decoration in the Russian palace.To modern people,it is visual fatigue and space cumbersome.All of this is want to present a very successful theme of Russian.

On the way to the reception hall.The "onion roof" shape will use to do the door frame.As well as the shape of the palace doors and windows.The crystal lamp will give it a golden outline.on the second floor.There is a white barrier for discussing areas.The light will express a gorgeous atmosphere.

Russia feeling: the simple meelow white and black

Othello floor plate design is a direct follow the example of St. Petersburg. Next summer palace garden in Othello disk pavement, black and white noble and elegant, showing a more golden royal honor.The white and black as the keynote of the interior design has become the best out of the golden background.

The most exciting part of the reception center is in the reception center of the axis.forms a fine pattern image of pure white wall.cylindrical geometry type stacked from floor to ceiling design is made of special tubes which are assembled one by one to human enormous project.There is no framework for coated and supported, And there is no paper tube between latches and Kan then attributed to the balance of force and gravity. The average spread, no variations to keep each perfect circle, the most environmentally friendly paper can be play out as the material effects of space, called basalt design creativity and initiative. White space with the introduction of natural light from the outside, like a noble and holy ritual space, the church of the Russian royal sense of space and contemporary design techniques to translate and less religious solemnity, still classic and elegant, but have become more contemporary impressionistic and poetic.

The black and white Russian is well-known Ruthless in the western. Russian vodka to be the basement, plus the coffee liqueur milk mixed with oil and lubricants into the cocktail.Sweet but strong momentum.It is like a wise bartender, the deep rich black, white, dense fine to reconcile Russia's magnificent Royal.and when the black and white be pure to the limit.That is another dazzling golden light which was even more profound.

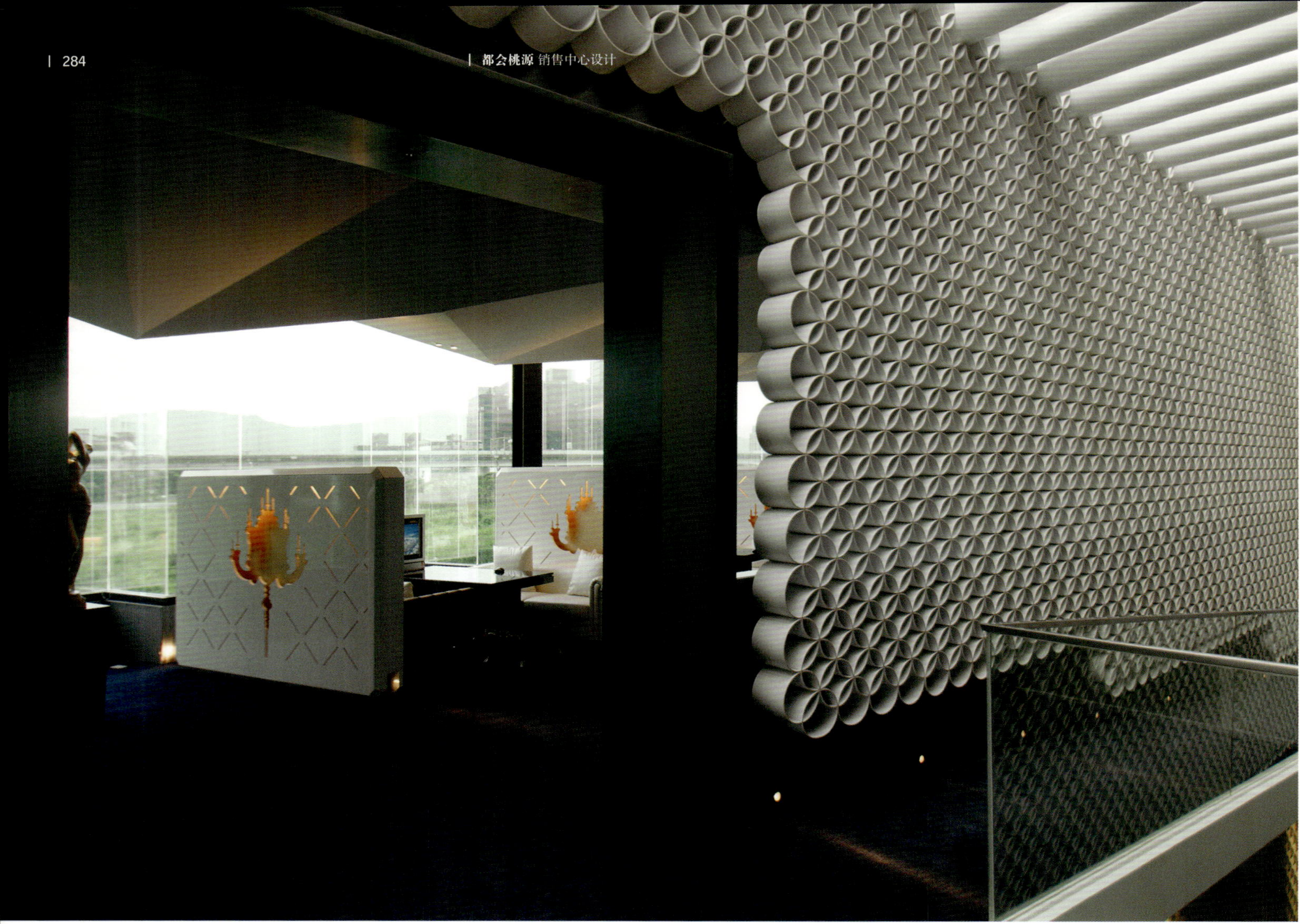

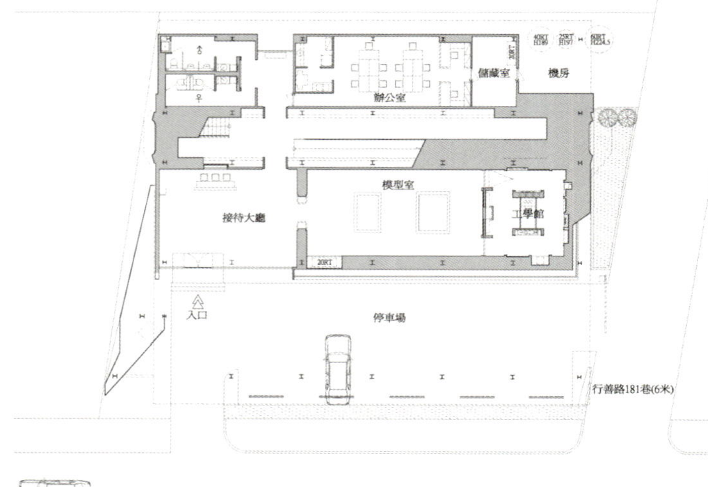

儲藏室
機房
辦公室
模型室
工學館
接待大廳
20RT
入口
停車場
行善路181巷(6米)

30RT
VIP
洽談區
30RT
樣品屋區

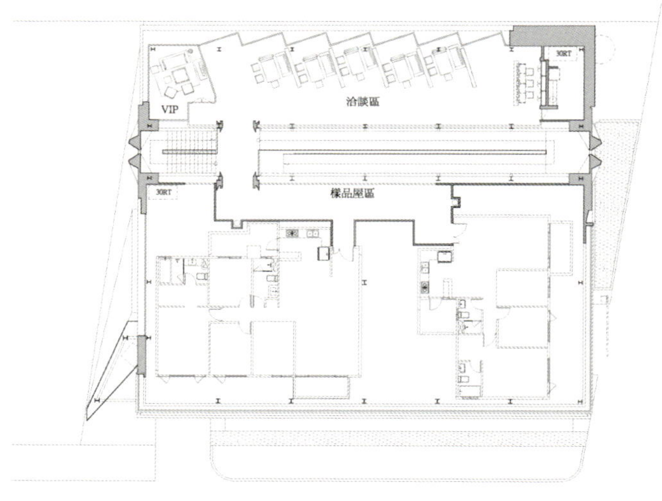

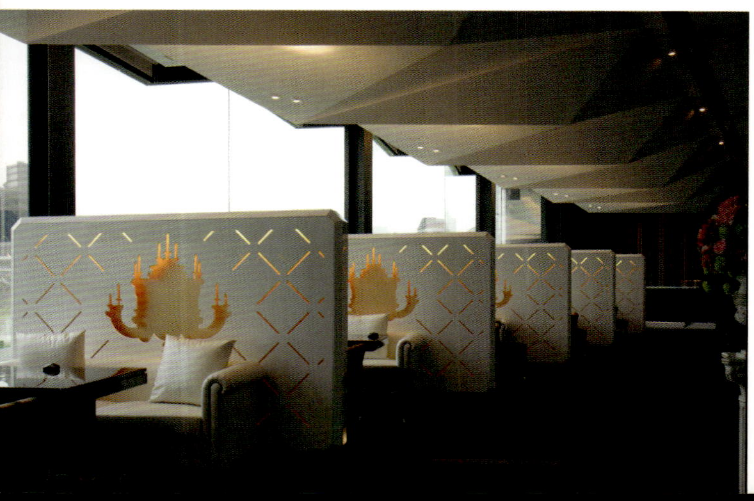

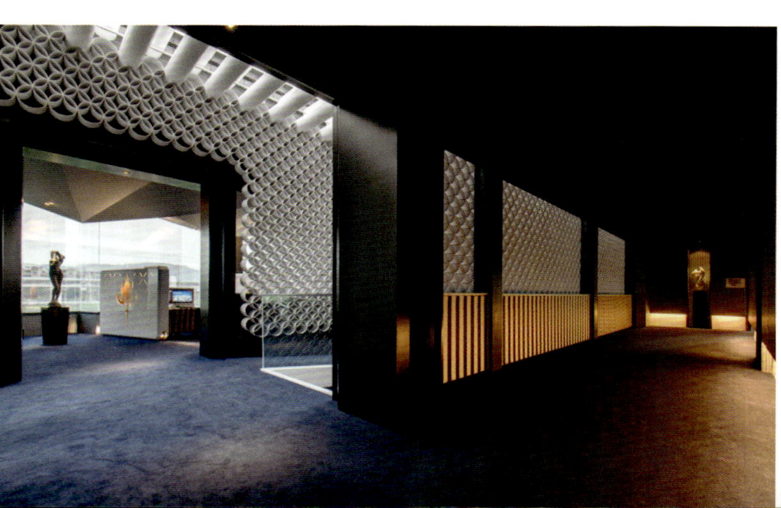

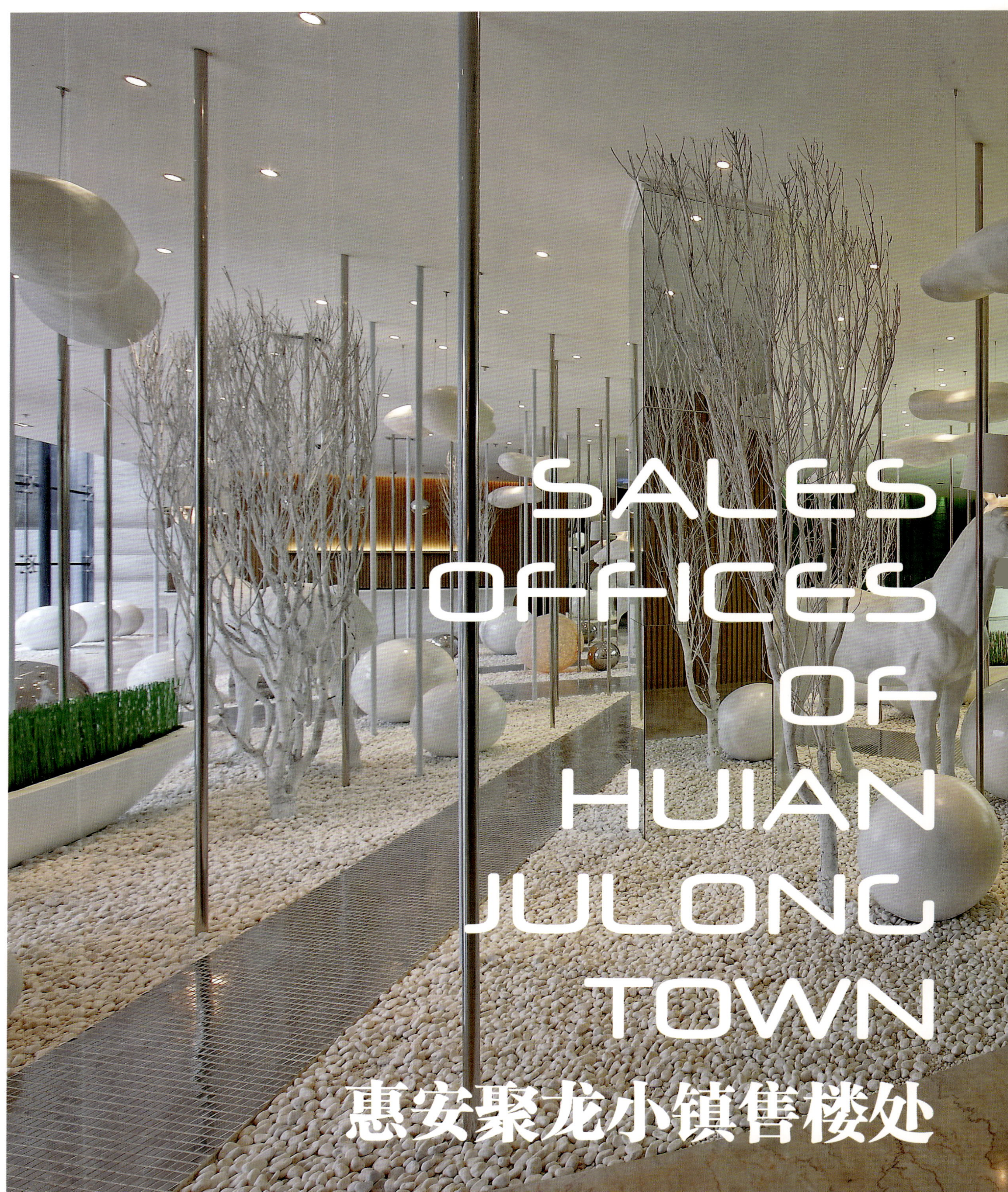

SALES OFFICES OF HUIAN JULONG TOWN

惠安聚龙小镇售楼处

翠绿环抱、白马仙居、云蒸雾绕……神仙眷侣的梦境空间。

本案是位于青山绿水间的商场建筑售楼处，只可惜设计空间无法延展到水淌风拂的户外，欲置写实造景于室内，反倒显得区区小景无法展现自然造化，只有升华成一种意境才能让顾客体悟到耳目一新的境界。因此，整体设计意在创造出一个飞离小景而氤氲出的意境，空间的每一个角落都渗透着浓厚稠密的梦境感。

入口由洁白的树枝、小石、缱绻云彩以及苍绿小舟引入清幽、净白的空间，诱人的阳光与白马、丛林的轻吻，让全身感官充满别有洞天的幸福之感。绿色墙体环抱而成的巨型沙盘区，青色圆管化作枝桠绵延飘浮在亮色的软膜天花之下，光影下透出丝丝的宁静于温暖。

洽谈区划成内外两个空间，外洽谈室由纯色通透方管枝条隔断围合；内洽谈区分为开放空间和半开放空间，适应不同客户的空间分流。

儿童娱乐室由循序渐进的红色墙纸以及大小各异的黄、橙色圆形壁龛组合，地面铺上如流水般拥有层次分明的的苍蓝色地毯，其上散放着青绿的圆形坐垫和可爱的布偶马，构成了一个魔幻的童话空间。

2000 m^2 售楼空间内，还有样品展示吧台，休闲区、梳妆区，办公室、财务、银行按揭室、员工餐厅、休息室等配套完备的功能性空间。在空间中穿越，视觉不停被升华的意境唤醒，身处其中，这份宁静清洗着疲惫的心灵，这份感动使人们不得不停下来，慢慢体味着这别致的意境空间。

项目名称：惠安聚龙小镇售楼处
设 计 师：赖云舟
项目地址：惠安
建筑面积：2000 m^2
主要材料：水曲柳、石材、绿草皮、不锈钢管、方管、白石子、白色枯树
设计单位：厦门大韵天成设计有限公司

Surrounded by green, cloud steaming fog around the ... ... dream space.

This case is located between the mountains of marketplace sales offices, but unfortunately the design space can not extend to the outdoor wind blowing the water drips, Yu Zhi realistic scenery in the room, hand, can not seem to show little scene mere natural good fortune, only sublimated into a mood to let the customer realized that a fresh state. Therefore, the overall design is intended to create a little scene and dense fly out of the mood, every corner of space are permeated with a strong sense of dense dreams.

There are white branches, stone, tempting clouds and the introduction of a green boat, also the quiet, clean and white space stay by the entrance.The attractive sun and white horse, so that the feeling of happiness is full of Journey. Surrounded by a giant wall which is made of green sand.Then it turned into a blue tube floating in the bright branches,it will give us the warm feeling.

Negotiation area was divided into two spaces.In the outside of the negotiating room by the permeability of the pipe branches.On the inside,it is also divided into two space which is complete openning space,the other one is the half opening space.Just for varies guests to communicate.

Children's playroom, and by the gradual red wallpaper sizes of yellow, orange circular niche mix, the ground covered with water, such as structured as the dark green with blue carpet, the chairs were on green cushions and cute round puppet horse, form a magical fairy tale space.

2000 sqm of sales space, and sample display bar, lounge area, dressing area, office, finance, mortgage banking rooms, staff restaurants, lounges and other supporting complete functional space. In space through the visual wake-up kept the mood to be sublimated, in which all, this quiet clean the tired soul, this move makes people stop and slowly savor the mood with this unique space.

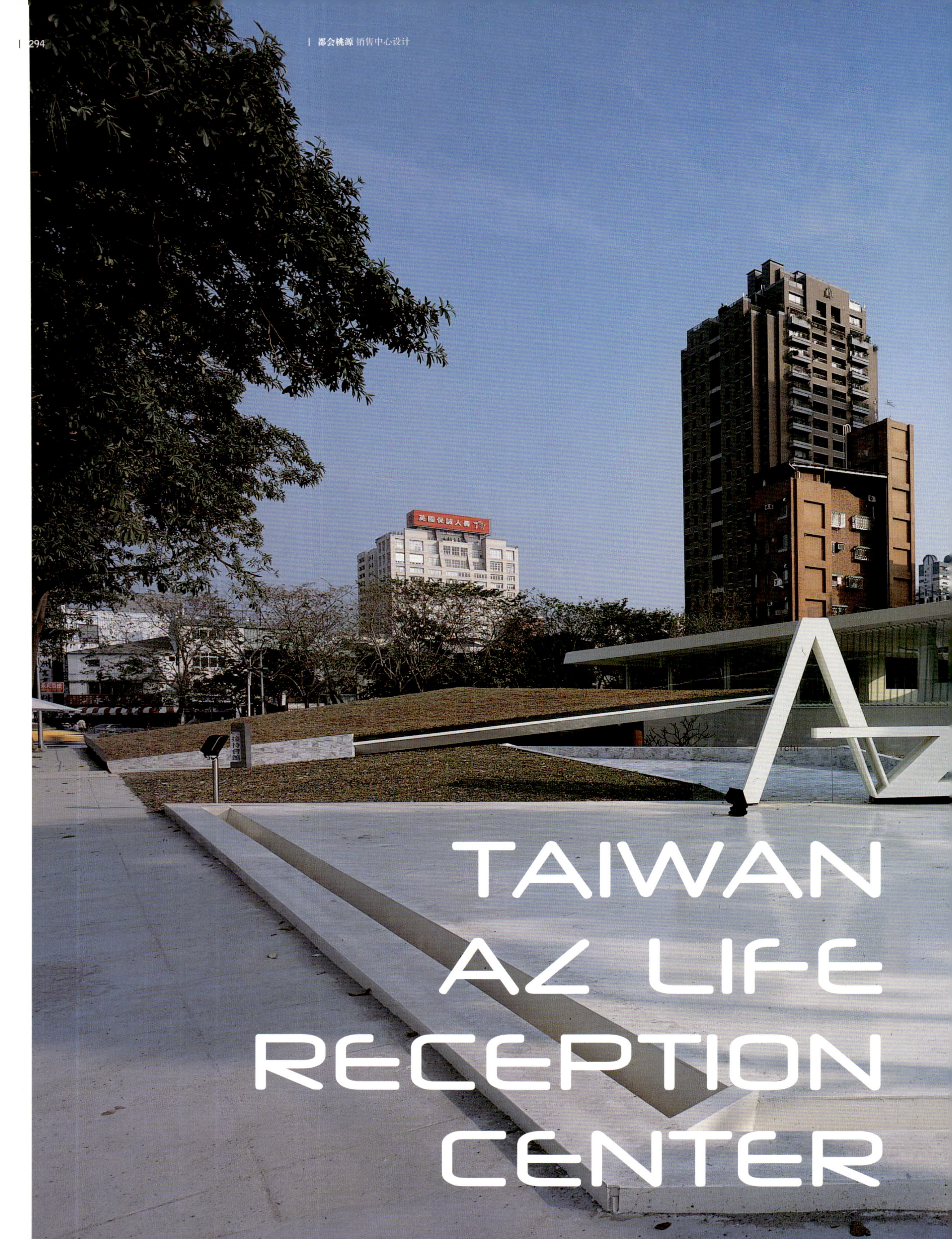

TAIWAN AZ LIFE RECEPTION CENTER

台湾 AZ 学生活接待中心

项目名称：台湾AZ学生活接待中心
设计公司：真工设计工程股份有限公司
设 计 师：程绍正韬
面　　积：826 m^2

基地位于台中市西区美村路与五权七街路，附近有美术馆、市立文化中心等配套，区域内文教气息浓厚，交通便利，向来是高端客户聚集的优质生活圈。"AZ学生活"定位以年轻族群为主力销售对象，宣传口号也非常吸引人：一栋视生活享受为圣域的精致住宅！在宣传中号称该项目是26味的华丽与禅意，是五觉的收纳，也是六感清静的韵律，总之非常具有品味与格调的小区。

设计师程绍正韬，精心研究富裕优雅的新世代族群之购买行为及心理特征，并以之为设计的基础，在面对美术馆的优质环境下，将接待中心的建筑刻意潜藏于都市广大绿环境的基地里，以Club Lounge的概念来取代传统之销售中心，创造时尚与创意的生活聚点；并在时尚与创新中，配合原始基地纹理，构建和谐之所在。

接待中心地块周边的浓荫绿树，几乎遮掩了这个别有洞天的处所，接待中心二层地面与街道平行，外部经过，只能看到二楼的白色墙身，如折纸雕塑一般，在阳光的照射下，绽放出异样的光彩。沿着基地边缘，长长的清水混凝土台阶走下，才发现在下沉式的广场里，藏着一个不受尘世干扰的清静之所。

下沉式广场的中心以一个菱形水池园

景与一个平台式的休闲广场构成。两边的建筑隔水相望，默然静立。繁忙的都市，因为这个下沉式的广场，难得地拥有了一方宁静恬然的空间。

走过长长的通道，来到洽谈区，黑与白的世界，优雅脱俗，品味超然。水体、玻璃构成的空间，轻灵通透，加上石材的装点与长毛地毯的铺垫，呈现出丰富的层次与质感。洽谈的过程也显得轻松从容。

想要全面了解项目，自然也要前往展示区进行多角度的观摩。从展示区首层沿着水泥通道转过石墙白壁，踏上台阶，来到二楼。二楼的通道豁然开朗，尺寸开阔的走廊，从地面到墙身到天花，甚至是柱体都一色净白，连墙侧白色石基花坛里种植的鲜花也是纯白的蝴蝶兰，到访者如同被空间净化了一般，宗教朝圣式的端严感油然而生。

透过这样的通道，再到样品屋，大气优雅的空间里，浓缩着东方文化的精气神，简约而宁静，优雅而质朴，成就了一个精致而清雅的居住空间。

如果说观念是对思想的洗礼，空间也可以完成对精神的净化，AZ学生活接待中心践行了这个设计理念。

AZ Life with base located at the intersection Meicun Rd and Wuquan Rd, Taizhong and facilities like Museum of Art, and Municipal Cultural Center, enjoys a strong flavor of culture and education, as well as convenient. As a good place for high-end customers who are in the pursuit of quality living, AZ Life targets young group as its potential purchaser; the slogan is very attractive: a sanctuary that takes life as an enjoyment. As alleged in the propaganda, the project is gorgeous and luxurious and full of Zen and rhythms to make your sensations clean and pure, totally a community of taste and style

Faced with such a qualified environment, Mr. Cheng, the designer pay great attention to the purchasing behavior and psychological characteristics of new generation of affluent populations elegant and taking it as his design basis, deliberately hides the connotation of the reception center into the vast urban green environment, replaces the traditional sales center with a concept of club lounge concept to replace, and thereby creates a space of stylish and creative, a harmonious one in line with the original base texture in fashion and innovation.

In the deep of the forest, the reception center seems to be invisible; up to its 2F, the building is parallel to the streets. Only from the white wall of the second floor, the building body like an origami sculpture can come into view, bursting out with a strange glory in sunshine. Along the base edge are long-concrete stairs, when coming down, you would suddenly find that such a quiet place is surprising to be built in a sunken plaza.

In the middle of the sunken plaza stands a diamond-shaped pool and a platform-style square. Silent buildings across the water are facing each other. Thanks to the plaza, the busy city provides people with a rare space, quiet and tranquil.

The negotiation area is at where the long corridor ends. That's a black and white world, refined elegant. The space of water and glass, is light, agile and transparent. Especially, the decoration of stones and the long-haired carpet bring out a wealth of richness and texture. Inquiry process can be in peace and at ease.

For a comprehensive understanding of the project, a multi-angle observation around the showcase is must. Through the concrete tunnel and behind the white stone wall, are the stairs to the second floor. The channel there suddenly become open and clear from the ground to wall to ceiling, and even whitening cylinder are of the same color. Along the white stone wall are white orchid flowers. It's a space that seems purified, and generates a pilgrimage-like sense.

Inside the sample house, there is an atmosphere of elegance and magnificence. There, the oriental culture, and the simple quiet sense makes here a living space of exquisite and refined.

If the idea is the baptism of thought, space is certain to be able to complete the purification of the spirit. That is what AZ Life maters; and that is what the ideas approached in this project.

古剛果
文物展
FROM RITUAL
CUTTING EDGE
童年夢工廠
~德國童話之旅
2005年1月29日-5月20日
威廉・莫里斯
與工藝美術特展
illiam Morris Arts & Crafts

SALES CENTER IN CHENGDU

成都金沙鹭岛售楼处

JINSHA LUDAO

密林中一片石墙斜穿，毛面、哑面、光面质感的黑色花岗岩混搭；一面水镜，几点星灯，几段木化石横卧，几阶木平台错落；为达至室内外相融，建筑的围护为无框钢化玻璃，支撑钢构屋顶的钢柱也尽量纤细，钢构屋顶仿佛漂浮在空中。接待台为花岗岩方料切凿，地灯映衬下自然断裂面的力度感更为强化；围绕着接待台的是几片横向延伸的木作承板，任意角度的黑钢斜片穿插其间。吧台上天花延伸而下的木格架罩着高低错落的云石光块；视听墙为简化的现代壁炉，呼应的是入口黑色石墙的质感组合；在大厅与后场之间，由斜墙延伸隔出的过道空间内，洗手台成为视觉焦点：一片镜面将椭圆形花岗岩方料一剖为二，门字形木格架又隔出几分私密；地面皇室木纹大理石从室内延伸至户外平台与天棚冲砂木的凹凸肌理相呼应。

Sales offices in the jungle, the facade with rough surface, matte, glossy black granite mix and match; mirage, spotlights, petrified wood, step together. In order to blend with the interior, the building envelope for the frameless glass, steel columns supporting steel roof is very small, the roof seems to float in the air. Front Desk for the granite material is more obvious against the background lights; reception desk is surrounded by a few pieces of wood, boards are placed at random. Between the hall and back, separated by a sloping wall extends out of the aisle space, sink into a visual focus: mirror oval granite stone will be split into two, a sort of split wood frame private; ground marble Extending from indoor to outdoor platform, and the echo texture of the ceiling.

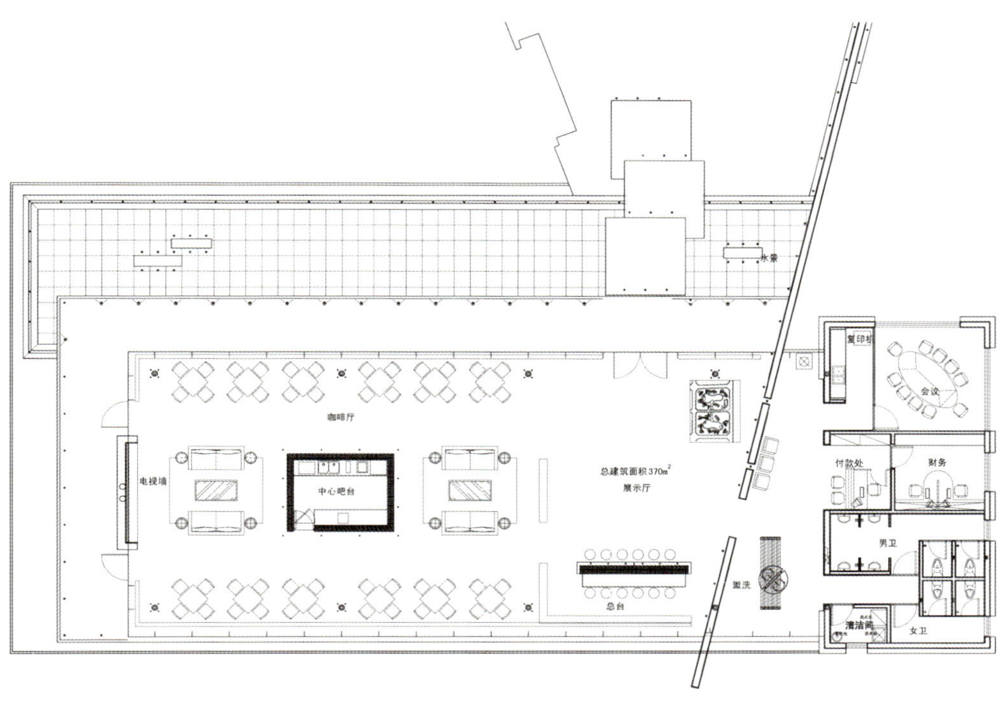

FANTASIA MIC PALZE SALES CENTER

花样年美年广场销售中心

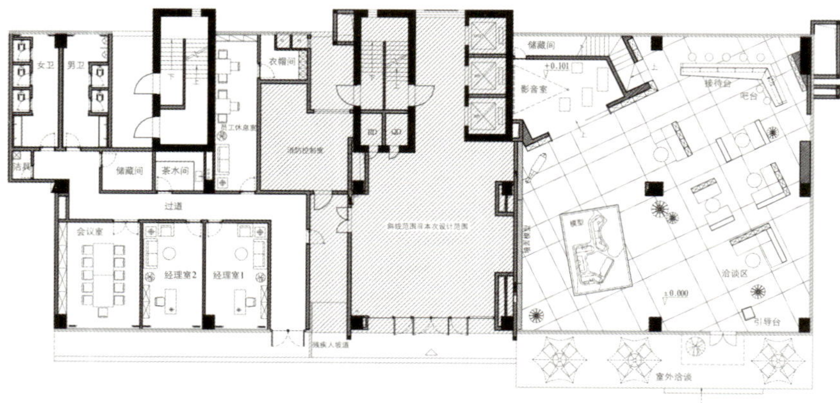
女卫
男卫
衣帽间
储藏间
影音室
接待台
吧台
员工休息室
消防控制室
洁具
储藏间
茶水间
过道
斜线范围非本次设计范围
会议室
经理室2
经理室1
模型
洽谈区
±0.000
+0.101
引导台
残疾人坡道
室外洽谈

作为一个有销售功能的咖啡吧，在平面分布时采用了较为活泼的形式。几组形式不同的沙发随意搭配，以及立体剪纸效果的树形灯，营造出梦幻的咖啡氛围。以“书”和趣味饰品为素材组成大面积的装饰表皮，出现在主立面，增加些许轻松随意的感觉。

项目名称：花样年美年广场销售中心
项目地址：深圳
项目面积：330 m^2
竣工时间：2010
主要材料：橡木饰面、人造石、白色镜面、编织地毯
设计单位：于强室内设计师事务所

As a sales function in planar distribution of coffee, used the relatively lively form. Optional collocation several groups of different forms of sofa, stereo paper-cut effect tree light, build a dream coffee atmosphere. With "book" and fun accessories for large adornment material composition, appears in the epidermis Lord facade, increase slightly relaxed and optional feeling.

图书在版编目（CIP）数据
都会桃源：销售中心设计：汉英对照 / 孔新民主编. -- 北京：
中国林业出版社, 2011.5
ISBN 978-7-5038-6160-4

Ⅰ. ①都… Ⅱ. ①孔… Ⅲ. ①销售—服务建筑—室内装饰设计—图集 Ⅳ. ①TU247.9-64

中国版本图书馆CIP数据核字(2011)第079145号

主　编：孔新民
策　划：纪　亮
编　写：孔新民　张　岩　王　超　刘　杰　孙　宇　李一茹
姜　琳　赵天一　李成伟　王琳琳　王为伟　李金斤
王明明　石　芳　王　博　徐　健　齐　碧　宋晓威
张文媛　陆　露　何海珍　刘　婕　夏　雪　王　娟
黄　丽　程艳平　高丽媚　汪三红　肖　聪　张雨来
陈书争　韩培培　付珊珊　张　雷　傅春元　邹艳明
高囡囡　杨微微　姚栋良　武　斌　陈　阳　张晓萌
魏明悦　佟　月　金　金　李琳琳　高寒丽　赵乃萍
裴明明　李　跃　金　楠　邵东梅　李　倩　左文超
陈　婧　陈圆圆　陈科深　吴宜泽　沈洪丹　韩秀夫
高晓欣　包玲利　郭海娇　阮秋艳　王　野　刘　洋
牟婷婷　朱　博　宁　爽　刘　帅
翻　译：牛晓霆

中国林业出版社 · 建筑与家居出版中心

出版联系：纪　亮　李　顺
联系电话：010-8322 3051
在线对话：1140437118（QQ）

出版：中国林业出版社
（100009 北京西城区德内大街刘海胡同 7 号）
网址：www.cfph.com.cn
E-mail：cfphz@public.bta.net.cn
电话：（010）8322 3051
发行：新华书店
印刷：恒美印务（广州）有限公司
版次：2011年5月第1版
印次：2011年5月第1次
开本：230mm × 300mm
印张：20.25
字数：200千字
定价：298.00（RMB）；60.00（USD）